RATES AND MECHANISMS OF CHEMICAL REACTIONS

RATES AND MECHANISMS

OF

CHEMICAL REACTIONS

W. C. Gardiner, Jr.

University of Texas

W. A. BENJAMIN, INC.
Menlo Park, California • Reading, Massachusetts
London • Amsterdam • Don Mills, Ontario • Sydney

Second printing, August 1972

ISBN 0-8053-3101-8
FGHIJKLMNO-AL-7987

Editor's Foreword

THOSE RESPONSIBLE for introducing physical chemistry to students must make difficult decisions in selecting the topics to be presented. Molecular physics and quantum mechanics have been added to the subject matter of physical chemistry in the past 35 years and are essential to the chemist's training. Yet many of the more classical areas of physical chemistry continue to be important, not only in chemistry, but also to an increasing extent in biology, geology, metallurgy, engineering, and medicine. Consequently there is pressure on teachers and textbook writers to cover more and more material, but the time available in the curriculum has not increased correspondingly, and there is a limit to the size of any textbook. Furthermore, it is difficult for any one author to write with authority about all of physical chemistry.

This text–monograph series is an attempt to make it easier to deal with this problem. The important basic topics of physical chemistry are covered at an introductory level in relatively brief, interrelated volumes. The volumes are written in such a way that if a topic of special interest to a student may not have been included in the course to which he has been exposed, he can learn about it through self-study. Consequently instructors can feel less reluctant to omit or condense material in their courses, and flexibility will be possible in the course plan, both from year to year and from institution to institution, in accordance with changing demands. The introductory presentation of physical chemistry in this form has the additional advantage that it permits a more detailed explanation of difficult points. It also permits the occasional inclusion of more advanced material to which the instructor can refer the more highly motivated students.

WALTER KAUZMANN

Princeton, New Jersey
March 1969

v

Authors' Foreword

PHYSICAL CHEMISTRY has been defined (by a practitioner) as "That part of chemistry which is fundamental, molecular, and interesting." Although this may be a useful guide in selecting a research problem, it is of very little help in planning an introductory course. Such a course should not be put together from a little bit of everything that has ever been called physical chemistry. Instead, it should concentrate on fundamentals so that wherever a student turns later, he can build on a secure foundation.

Physical chemistry has the general task of explaining the causes of chemical behavior. The essential, irreducible fundamentals of the subject are four in number:

1. Quantum mechanics: the mechanics of atoms and of their combinations in molecules;
2. Statistical mechanics: the framework by which molecular properties can be related to the macroscopic behavior of chemical substances;
3. Thermodynamics: the study of energy and order–disorder, and their connections with chemical changes and chemical equilibrium;
4. Kinetics: the study of the rates of chemical reactions and of the molecular processes by which reactions occur.

Many additional topics are found in introductory physical chemistry textbooks. These include methods of molecular structure determination, the several branches of spectroscopy, electrochemistry, surface chemistry, macromolecules, photochemistry, nuclear and radiation chemistry, and theories of condensed phases. These are essentially applications of the fundamental concepts, and in our books they are taught as such. The relative emphasis given to these topics will vary with the nature and level of the course, the needs of the students, and the inclination of the instructor. Yet a secure grounding in the four fundamentals will give a chemist not everything that he needs to know in physical chemistry, but the ability to recognize and learn what he needs to know as circumstances arise.

These books are an outgrowth of our experience in teaching the basic physical chemistry course over the past few years at the Universities of Illinois, Colorado, and Texas, and the California Institute of Technology. Each of us has written the part of the course that he knows best. Hopefully this approach will avoid the arid style of a book-by-committee, and yet allow each topic to be covered by an author who is vitally concerned with it. Although the present order—quantum mechanics, thermodynamics, kinetics—appears to us to be the most desirable pedagogically and the most obviously unified to the student, the material has been written so thermodynamics can precede quantum mechanics as in the more traditional course.

The authors feel strongly that basic physical chemistry should be presented as a unified whole rather than as a collection of disparate and difficult topics. We hope that our books will serve that end.

MELVIN W. HANNA
RICHARD E. DICKERSON
W. C. GARDINER, JR.

March 1969

Preface

REACTION MECHANISMS and the theories of elementary reaction rates are of interest and importance to virtually all chemists and chemical engineers. The traditional place of reaction kinetics in the junior physical chemistry course has nonetheless eroded considerably in recent years, mostly to allow time for introducing students to quantum mechanics at an early stage in their studies. As the time available to teach kinetics was shrinking, practicing kineticists were making substantial additions to the core of fundamental knowledge about reaction kinetics of which all chemists and chemical engineers should be aware. The net result is that there is more reaction kinetics that ought to be taught and less time in which to teach it. For this reason a kinetics textbook now has to meet sharply conflicting demands. On the one hand, it should be short and readily assimilable in order that the material can be learned in the available time. On the other hand, it should also have the flavor and scope of a comprehensive monograph, since a chemistry student can progress from his first physical chemistry course all the way to the PhD degree without ever again hearing about kinetics in the classroom and may want to refer back to his undergraduate textbook later in order to orient himself concerning some aspect of the subject.

This book is written in the spirit of a compromise favoring an ambitious student who seeks to attain, in the most direct way, a working knowledge of the basic principles of experimental chemical kinetics, and then to study further the fundamentals of whatever special topics he finds interesting. In making this compromise, it was expedient to treat some topics in far greater depth than others. The section on heterogeneous catalysis, for example, is quite superficial, while the theory of unimolecular reactions is developed in extensive detail.

Since this is a textbook rather than a monograph, there are no footnote references to the research literature and no detailed descriptions of experimental methods. These are of interest to instructors, but more a distraction than a help to beginning students. Those who are uncertain as to whether the material is up to date can satisfy themselves by comparing it to the monographs and literature references that are cited as supplementary references at the end of each chapter. The supplementary references will also prove useful to readers who wish to inform themselves further about topics of special interest to them.

Many important ideas are developed in the exercises within the text. No reader should omit doing them. The exercises at the ends of some chapters reinforce and extend the subject matter, and are intended primarily for readers who want to advance beyond the introductory level. Anyone who does all 141 exercises will have attained an excellent command of the subject at an advanced level.

Individual courses will vary widely in the amount of attention devoted to each part of this text. A recommended core would consist of a thorough treatment of Chapters 1, 2, and 3, and Sections 4-1, 4-2, and 4-7. This could be followed by brief discussions, or reading assignments, in the rest of Chapter 4, and Chapters 6, 7, and 8.

There is more than enough material in this book for a one semester senior or graduate course. Most instructors using it for that purpose, however, will want to supplement it in one place or another.

The manuscript of this book was improved by comments from colleagues and students too numerous to mention and especially by the detailed criticism given the final manuscript by Professor Walter Kauzmann; without the ideas and discoveries of the many contemporary chemists whose work fills these pages it would not exist at all.

W. C. GARDINER, JR.

Austin, Texas
February 1969

Contents

Chapter 1

INTRODUCTION TO CHEMICAL

KINETICS

CHEMICAL REACTION is a subtle, complex natural phenomenon that chemists are only beginning to understand. A body of detailed, molecular-scale knowledge about chemical reactions comparable to present knowledge of molecular structures does not yet exist; the theories of chemical reactions are not yet as well developed as the quantum mechanical theory of molecular structures. This book describes how chemists do experiments and devise theories to broaden and deepen what understanding of reactions there is. In reading it you will find that the relevant experiments and theories have mostly to do with the *rates* of chemical reactions; for this reason the name *chemical kinetics* has come to denote the area of science concerned with the study of chemical reactions. If you should find that chemical kinetics is an underdeveloped science compared with other aspects of chemistry, be tolerant and recognize that time-dependent problems are intrinsically more difficult than equilibrium ones, or be challenged and spend some of your scientific lifetime improving the situation. In either case, it is important for you as a chemist to learn what the questions of chemical kinetics are, even though you may not live long enough to find what the answers are.

Chemical kinetics has three levels of meaning. First, chemists or engineers who must develop efficient synthetic procedures for practical purposes need to know the factors that influence the rates of their reactions in order to proceed rationally with their work. We shall discuss the experimental and mathematical methods that provide starting points for solving such practical problems. A second level of meaning, to which we devote most of this book, has to do with deducing molecular descriptions of chemical reactions from reaction rate measurements and other observations. At this level we have not really left practical problems behind, because here, as in many kinds of scientific research, more detailed understanding often leads to new and sometimes to useful discoveries. At this level of meaning we recognize a large segment of descriptive chemistry, for here the subject of *reactivity* emerges as the body of generalizations about which chemical transformations can occur. Beyond this level of meaning we come to the final one, where chemists attempt to understand how individual molecular transformations occur. They wonder about the stereochemistry and the electrons of molecules undergoing reactions in refluxing mixtures in flasks; about the flow of energy within an isolated molecule about to fly to pieces; about the different stages of binding to a solid catalyst surface; about the motions of ligands being exchanged at an ion; and about countless other details of molecular change.

To illustrate these three levels of meaning let us consider the chemical reaction

$$H_2O + CO \rightleftharpoons CO_2 + H_2$$

The rate of this reaction is important in determining the CO/CO_2 and H_2O/H_2 ratios in the exhaust gases of Bunsen burner flames, internal combustion engines, jet aircraft engines, and rockets. Chemists or chemical engineers interested in optimizing the power output of hydrocarbon fueled engines, or in reducing their carbon monoxide output to suppress air pollution, need to know the rate of equilibration of this reaction.

Underlying the practical importance of the equilibration rate is the question of how this equilibrium is maintained at the molecular level. The answer is that the reaction is catalyzed by hydrogen atoms and hydroxyl radicals according to the reactions

$$H_2O + H \rightleftharpoons H_2 + OH$$

$$OH + CO \rightleftharpoons CO_2 + H$$

These two reactions represent what happens at the molecular level, whereas the chemical reaction itself only represents the macroscopic observation that water and carbon monoxide are converted to carbon dioxide and hydrogen. If the two catalytic reactions are each at equilibrium, then their sum, which is the observed chemical reaction, is also at equilibrium. Our second level of meaning is therefore a clarification at the molecular level of the way a chemical conversion occurs. Underlying this clarification at the molecular level we have the question of how atomic rearrangements occur when three hydrogen atoms and an oxygen atom—or two oxygen atoms, a carbon atom and a hydrogen atom—are temporarily close together. We would like to know the evolution of the chemical bonds while one arrangement of atoms (e.g., $OH + CO$) is converted to another, that is, $CO_2 + H$. Our third level of meaning of chemical kinetics is, in this example, an inquiry into the atomic positions and motions during the conversion of OH and CO reactant molecules into the products CO_2 and H.

For some more introductory examples of the questions asked in chemical kinetics, let us consider the synthesis of sugar. Eight generations of chemists have known that green plants convert six moles of carbon dioxide and six moles of water into one mole of a hexose and six moles of oxygen. On the level of atoms and molecules, how does this come about? Through what series of encounters and molecular configurations do atoms from water and carbon dioxide molecules wander before they appear as sugar? Why those encounters rather than others? How long do the encounters last? What happens to the electrons during the encounters? What is it about the enzymes which catalyze these reactions that makes the encounters shuffle atoms so specifically?

Or let us consider pure water. A high school chemistry student can tell you that water has pH 7 and ion product 14. He can also tell you the reaction that occurs when acids and bases combine to form water. But how do hydroxyl and hydrogen ions actually reach one another through their hydration shells to disappear as water molecules? How do the structural reorganizations, the making and breaking of hydrogen bonds, and the electron transfers occur? Here we have the molecular processes that underlie pH 7 and ion product 14.

These are samples of questions that concern chemical kinetics. The answers that now can be given to these questions and to others like them are partial, tentative, and approximate.

Let us begin our study of how chemists seek these answers by reviewing the thermodynamics of chemical equilibrium and by introducing some terms used to describe the ways in which systems achieve chemical equilibrium.

1-1 CHEMICAL EQUILIBRIUM

Formulas derived from the laws of thermodynamics, taken together with measured thermal properties of elements and compounds, provide the means for direct calculation of the equilibrium chemical composition of any system. Given the percentages of those elements that comprise the system, and knowledge of the compounds that can be formed among them, we can calculate this equilibrium composition for any values of temperature, density, and so on, of interest. If many elements are present, this may be a long and tedious calculation, perhaps best done with the aid of a large electronic computer. But regardless of the complexity of the equilibrium composition or the values of the temperature or other variables, the equilibrium composition is determined in principle by thermodynamic laws and calculable in practice if the required thermal data are available. Specifying the values of two state variables is usually sufficient to define the equilibrium state of a system containing a fixed quantity of matter. For each choice of two such variables there is a corresponding thermodynamic function, called the *characteristic function*, or the *natural function*, that has an extremum value at equilibrium. For a system constrained to constant temperature and pressure, the Gibbs free energy G has its minimum value at equilibrium; for constant temperature and volume, the Helmholtz free energy (or work function) A has its minimum value at equilibrium; for constant energy and volume, the entropy S has its maximum value at equilibrium.

For thermodynamic contemplations it is convenient to consider extremum values of characteristic functions as the logical and the best criteria for determining whether a given system is in an equilibrium state. For practical applications, however, the thermodynamic equilibrium criteria must be utilized quite indirectly. This is due to the form of the thermodynamic equations: if the constraints are directly measurable quantities, the characteristic function is not, whereas if the characteristic function is measurable, one or more of the constraints is not. A more practical, but less reliable, indication of equilibrium is that the system does not change with time. Some chemical intuition may be needed to avoid mistakes in judging whether a lack of observable change is an indication of chemical equilibration. This book, for example, is not in equilibrium with its environment. (If you are in doubt, touch a match to it and observe the change.) On the other hand, cessation of precipitation of silver chloride in the classic analysis for chloride ions can be taken as a satisfactory indication that equilibrium has been reached. So *time* intrudes, in this practical view, upon thermodynamics. We shall find this intrusion to be reasonable, in fact unavoidable. In applying thermodynamics to chemical reactions, the time scale of the observer, limited by the speed of his fastest instruments and by his patience or lifespan, must be considered.

For studying chemical reactions it is necessary to study systems that are not at chemical equilibrium. However, for two reasons it is valuable to know what the equilibrium composition will be when equilibrium is reached: the first is so the experimenter can identify the reaction or reactions that can possibly occur; the second is so he can determine whether attainment of equilibrium will lead to changes in the properties of the system large enough to be observed. An important instance of the first reason is identification of the *direction* of chemical change in a system in which the possible changes are not complicated. The only changes that ever occur are ones that bring the value of the characteristic function nearer to its extremum value. In a system constrained to constant pressure and temperature, for example, the only possible reactions are ones that decrease the value of the Gibbs free energy; ones that would increase it never occur.

The approach to chemical equilibrium can be studied in two ways. One way is to start with a system that is not at equilibrium and observe the equilibration while leaving the constraints (e.g., the temperature and pressure) fixed. This is the traditional and more generally useful experimental method in chemical kinetics, and the one to which we devote most of our attention in this book. The other way is to start with a system at equilibrium, perturb it by altering the value of one of the constraints, and observe the reestablishment of equilibrium. This method, called *relaxation spectroscopy*, is particularly useful for the study of rapid reactions. We discuss this approach in Chapter 9.

1-2 REACTION MECHANISMS

Chemical reactions are introduced to chemistry students in terms of chemical equations, such as

$$PbS + 4H_2O_2 \rightarrow PbSO_4 + 4H_2O$$

The stoichiometric character of chemical reactions is expressed by the coefficients used to balance such chemical equations. In simple cases, the equilibrium composition of a system in which only one reaction occurs can correspond to almost complete conversion of reactants to products. These are the cases in which the stoichiometric coefficients and the molecular weights of the compounds involved in the reaction suffice to calculate weight relationships among the various reactants and products. The chemical equations of reactions that are said to "go to completion" can be given a very simple interpretation. In the example given this interpretation would be as follows: one mole of lead sulfide reacts with four moles of hydrogen peroxide to produce one mole of lead sulfate and four moles of water.

The rearrangement of elements into different compounds, expressed by a chemical equation, is of course a laboratory-scale manifestation of the fact that atoms have been rearranged into different molecules. You might read the above chemical equation as: one lead sulfide molecule and four hydrogen peroxide molecules react to produce one lead sulfate molecule and four water molecules. But is this really what we want to say? Do the five reactant molecules actually convene, shuffle atoms, and depart as product molecules? Try to draw a picture of such a convention: better still, try to imagine such a convention for the reactions

$$C_8H_{18} + 12\tfrac{1}{2}O_2 \rightarrow 8CO_2 + 9H_2O$$

or

$$3Cu + 8HNO_3 \rightarrow 3Cu(NO_3)_2 + 2NO + 4H_2O$$

Should we interpret chemical equations as representing molecular occurrences? Obviously we should not.

Nevertheless, we are sure that the chemical reaction expressed by the chemical equation does reflect molecular events, and that reactant molecules are indeed converted to product molecules. Furthermore, it is plausible to assume that the events at the molecular scale are some kind of molecular meetings at which atoms are rearranged. We are led in this way to the notion of the *mechanism* of a chemical reaction as *the set of molecular events that results in the observed conversion of reactants to products.* We call the events themselves *elementary reactions*, or sometimes *steps*, and write them down just like we write down chemical equations. For example, there is a chemical reaction whose *chemical equation* is

$$2N_2O_5 \rightarrow 2N_2O_4 + O_2$$

It is believed to have a *mechanism* comprised of the following *elementary reactions* (the *molecular events!*)

$$N_2O_5 \rightarrow NO_2 + NO_3$$

$$NO_2 + NO_3 \rightarrow N_2O_5$$

$$NO_2 + NO_3 \rightarrow NO_2 + O_2 + NO$$

$$NO + N_2O_5 \rightarrow 3NO_2$$

$$NO_2 + NO_2 \rightarrow N_2O_4$$

Note carefully that equations which represent the stoichiometry of a chemical reaction and equations which represent elementary reactions look exactly alike but have entirely different meanings. It would be a simple matter for us to distinguish between the two in this book, for example by writing different kinds of arrows. We refrain from doing so not to avoid innovation but to put the reader on guard that he may encounter chemists who fail to distinguish the two meanings in their thinking as well as in their equations.

This book is about two aspects of mechanisms. The first is discovering them: what experiments can be done to find the mechanisms of particular reactions or types of reactions; what are the pertinent data, and how should they be analyzed; how can we decide among alternative combinations of elementary reactions; what mechanisms have been found for different kinds of chemical reactions?

The second aspect of mechanisms concerns the individual elementary reactions. Since these are molecular events, they should and do have theoretical interpretations based in various ways on our knowledge about molecules that are not undergoing reaction and upon hypotheses about what happens during elementary reactions.

In the first half of this book we deal more or less separately with finding mechanisms (Chapters 2 and 3) and with theories of elementary reactions (Chapters 4 and 5); in the second half of the book experimental results and theoretical interpretations are presented together.

"Finding the mechanism" is an overly simple phrase for describing the complex mental activity that intervenes between laboratory experiments and a proposed identification of the molecular events responsible for the chemical reaction under study. Mechanisms arise through inductive, rather than deductive, thinking. This forces chemists who study reaction mechanisms into a sometimes uncomfortable situation. On the one hand, it is important for them to know in addition to the rules introduced in Chapters 2 and 3 a large number of facts, namely, the results and mechanistic interpretations of all chemical kinetics studies related to the reaction they themselves are studying. This enables them to profit from the results of previous experience and ensure that their mechanisms do not have elementary reactions that imply conflicts with previous experimental results. On the other hand, it is also important for them to use a large amount of imagination to find mechanisms that imply their experimental results. The mind of the chemist studying mechanisms should, therefore, be both full and open at the same time.

It is important to remember that although mechanisms logically imply results, they must be found in the first place by some nonlogical process combining experience, intuition, and guesswork.

The theoretical description of elementary reactions has proved to be far more difficult than the quantum mechanical description of molecules themselves or the thermodynamic and statistical mechanical description of matter at equilibrium. We shall find the theories of elementary reactions to be correspondingly less satisfactory than thermodynamics, quantum mechanics, or statistical mechanics.

Let us introduce the term *encounter* for the circumstance that two molecules, or for some elementary reactions three molecules, are temporarily close enough together that electron motions are perturbed. Usually only a small fraction of encounters turns out to involve rearrangement of atoms, that is, reaction. Most encounters are merely elastic collisions, except for changes in rotational quantum numbers. We can describe the task of the chemical kinetics theoretician quite simply. He must face these questions: What fraction of encounters involves chemical reaction? What relative orientations of the molecules approaching one another and what internal and relative motion prerequisites distinguish reactive encounters from the others? What are the characteristics of the reactive encounter at the instant of maximum interactions among all the electrons, which we shall call the *transition state*? In what quantum states do the product molecules depart from the encounter? If the encounter is such that one of the molecules departing from it subsequently undergoes an atomic rearrangement without further encounters, how much time elapses between the encounter and the rearrangement, and how does the rearrangement come about? In Chapters 4 and 5 we discuss some answers that can be given to these questions.

By our definition of reaction mechanism, a chemical reaction has been given a mechanism as soon as a set of elementary steps has been written down. However, more detail than this is sometimes implied. In organic chemistry, the molecular conformations and the shifts of bonding electrons in the transition state are considered part of the mechanism. The context should tell if such wider meanings of the term mechanism are intended.

Some mechanisms are simple, others very complex; for distinguishing between simple ones and complex ones let us take the rule that the elementary steps of a simple mechanism can be counted on one hand. The simplest mechanism of all is obviously one with only a single step. The Diels–Alder reaction forming cyclohexene from ethylene and butadiene is believed to have

$$CH_2=CH_2 \quad + \quad CH_2=CH-CH=CH_2 \quad \longrightarrow \quad \text{(cyclohexene)}$$

a mechanism identical to its chemical equation. As an extreme example of a complex mechanism, in the reaction between hydrogen and oxygen the observed products are water and, under some conditions, traces of hydrogen peroxide and ozone. At least 26 elementary reactions, however, are included in the mechanism. From these two examples we see that a simple chemical reaction may have a complex mechanism, whereas a complex chemical reaction may have a simple mechanism.

It is a strict requirement for any mechanism that the net effect of the elementary reactions must be the chemical reaction concerned. In a simple mechanism, the net effect can usually be demonstrated algebraically by adding together the elementary reactions, analogous to solving a Hess' law problem in thermochemistry. Thus, when nitrous oxide is heated it decomposes to form nitrogen and oxygen

$$2\,N_2O \rightarrow 2\,N_2 + O_2$$

The mechanism may be the two elementary reactions

$$N_2O \rightarrow N_2 + O$$

$$O + N_2O \rightarrow N_2 + O_2$$

The sum of these two elementary reactions is the chemical equation for the decomposition. The situation becomes more complicated when the decomposition is carried out at higher temperatures, for then nitric oxide is also found as a product. Defining y to be the number of moles of O_2 formed for each mole of N_2O decomposed, the stoichiometric equation for high temperature conditions may be written

$$N_2O \rightarrow y\, O_2 + (y + \tfrac{1}{2})\, N_2 + (1 - 2y)\, NO$$

A third elementary reaction can be proposed to explain the formation of NO

$$O + N_2O \rightarrow 2\, NO$$

If the three elementary reactions are multiplied by $\tfrac{1}{2}$, y and $(\tfrac{1}{2} - y)$ respectively, their sum is once again the stoichiometric equation; the ratio $y/(\tfrac{1}{2} - y)$ emerges as the relative probability for an $O + N_2O$ encounter yielding $N_2 + O_2$ rather than 2 NO. For mechanisms more complex than this (such as the chain reactions that we shall study in Chapter 8) showing algebraically that the elementary reactions have the net effect of the chemical reaction may be extremely involved and not at all instructive. Nonetheless it is worth bearing in mind that every mechanism must imply the observed reaction qualitatively, perhaps even especially for reactions with many products, even though it may be a fruitless endeavor to find the exact linear combination of the elementary reactions that implies the observed product distribution quantitatively.

Catalysis has a simple interpretation in terms of reaction mechanism. Addition of a catalyst speeds up, or alters the direction of, a chemical reaction by introducing the possibility of new elementary reactions. Addition of iron salts to aqueous hydrogen peroxide catalyzes the decomposition reaction

$$H_2O_2 \rightarrow H_2O + \tfrac{1}{2} O_2$$

by providing the new elementary reactions

$$Fe^{3+} + HO_2 \rightarrow Fe^{2+} + O_2 + H^+$$

$$Fe^{3+} + HO_2^{\;-} \rightarrow Fe^{2+} + HO_2$$

$$Fe^{3+} + O_2^{\;-} \rightarrow Fe^{2+} + O_2$$

$$Fe^{2+} + H_2O_2 \rightarrow Fe^{3+} + OH^- + OH$$

$$Fe^{2+} + OH \rightarrow Fe^{3+} + OH^-$$

which in addition to the acid–base equilibria

$$H_2O_2 \rightleftharpoons H^+ + HO_2^-$$

$$HO_2^- \rightleftharpoons H^+ + O_2^{2-}$$

$$H_2O \rightleftharpoons H^+ + OH^-$$

lead to rapid decomposition.   Finding a set of coefficients that can be used to obtain the stoichiometric equation as a sum of these elementary reactions is left as an exercise for the student.

Let us conclude these introductory observations about mechanisms with a small dose of skepticism: Reaction mechanisms are *provisional, hypothetical suggestions* based upon available experiments, and do *not* necessarily represent the way in which reactions *actually* happen.   We require that the implications of any proposed mechanism agree with the experimental facts.   It is not possible to go farther than this.   If the mechanism of some reaction is said to be well established, you should regard this only as a way of saying that the mechanism is compatible with the data taken and similar to other mechanisms compatible with data for similar reactions.

The notions introduced in this section will become clearer when we discuss specific mechanisms in detail.

1-3   REACTION RATES

The most important experiments for studying reaction mechanisms turn out to be experiments concerned with the rates of chemical reaction, that is, the quantity of reactant(s) converted into product(s) per unit time.   The central idea is simple.   If the frequency of encounters between molecules that *might* react with one another is changed, then so is the frequency of encounters that *do* lead to reaction; consequently, the measured conversion rate changes also. Higher concentrations of reactant molecules lead to more frequent encounters; more frequent encounters lead to more rapid conversion.   To take a specific example, H atoms in a stream of argon will form HO_2 if they are added to a stream of O_2.   The conversion rate is found to be proportional to the concentrations of H, Ar, and O_2.   From this fact it may be concluded that the reactive encounter involves all three species

$$H + O_2 + Ar \rightarrow HO_2 + Ar$$

and that the rate of conversion is governed by the rate of this elementary reaction.   Similar kinds of conclusions underlie the interpretation of all experiments in which mechanisms are studied by measuring reaction rates.

The experimenter observes the effects of concentration changes on conversion rates in order to see what kinds of encounters are involved. Even for simple mechanisms, however, it is sometimes difficult to deduce from assumptions about the elementary reactions exactly what the dependence of reaction rate upon reactant concentrations should be. Moreover, the immediate task facing the experimenter (or the student) is of course the opposite: to discover a mechanism using data on the variation of conversion rate with reactant concentration, and this is likely to be still more difficult.

The rate of an elementary reaction depends on two things only: the frequency of encounters of the kind described by the left-hand side of its reaction equation, and the fraction of encounters that are reactive. In experiments done at constant temperature, the encounter frequency is directly proportional to the concentration of each species represented. For individual elementary reactions in any mechanism, therefore, the dependence of rate on concentrations is immediately known. *Unimolecular* elementary reactions, with one species on the left-hand side of the reaction equation, have rates proportional to one concentration; *bimolecular* elementary reactions have rates proportional to the product of two concentrations, or, if the encounter is between two identical molecules, to the square of one concentration; *termolecular* elementary reactions have rates proportional to the product of three concentrations, to the product of one concentration times the square of another concentration, or to the cube of one concentration.

Rates of elementary reactions are seldom directly measured; the experimental data usually reveal how the rate of the overall reaction is affected by changing concentrations. As a preview of the kind of thinking we shall need in order to analyze experimental data, let us consider the following simple example.

Suppose that we were studying, at constant temperature, a chemical reaction

$$A + D \rightarrow products$$

having the mechanism

$$A + B \rightarrow C \tag{1}$$

$$C \rightarrow A + B \tag{-1}$$

$$C + D \rightarrow products + B \tag{2}$$

Here B is some catalyst, and C is presumed to be present only in very low concentration. It might be that reaction (1) and its reverse are very fast compared with reaction (2). Then compounds A, B, and C would be almost at equilibrium, and the concentration of C would be $[C] \simeq K_{eq} [A][B]$. The rate of product formation, which is proportional to the number of C–D encounters, would then be proportional to the product of the concentrations of A, B, and D.   On the other hand, it might be that reaction (2) is very fast compared to (1).   In this case, reactive C–D encounters can convert C into products as fast as C is produced in reactive A–B encounters, and the rate of product formation would then be proportional to the number of A–B encounters only, or the product [A][B].   The experimenter who finds the latter situation to prevail might test the mechanism by making [D] very small in order to reduce the rate of (2), thereby allowing (-1), the reverse of (1), to become appreciable.   This would restore the first situation, and it should then be found that the measured conversion rate is also proportional to the lowered [D].

Most chemical reactions have mechanisms with a large number of elementary reactions, and it may become a challenge to decide how concentration changes will affect the measured rates.   To make matters worse, most mechanisms assume the participation of elementary reactions involving species that are unknown as stable molecules; they may be so reactive that their concentrations never rise to observable levels. Once again, it is the inverse problem that is actually confronted: to devise a mechanism consistent with the rate, and other, data.

In later chapters we consider these matters in detail.   We anticipate now that a chemist studying a mechanism by rate measurements has a threefold task: first to find suitable ways to measure the conversion rate, then to find how it changes when concentrations are changed, and finally to devise and test further a mechanism consistent with his experimental data.   We shall find that measuring the temperature dependence of the conversion rate, borrowing from experience with other reactions, and observing the effects upon the conversion rate of adding other reactants prove to be helpful too.

1-4   OTHER MEANS OF STUDYING MECHANISMS

Most of this book is about the use of conversion rate experiments to study reaction mechanisms.   Rate measurements by themselves, or in combination with knowledge of appropriate thermochemical properties of the molecules involved, may be sufficient for obtaining satisfactory understanding of a reaction mechanism.   However, other kinds of experiments often provide essential additional information.   We survey here the most important ones; throughout the rest of the book we shall encounter examples of their use.

Product analysis and stereochemistry.   Chemical analysis is difficult work; if a mixture of minute quantities of similar substances is to be analyzed, it is very difficult.   For studying reaction mechanisms, however, detailed analysis of the reaction products may also be quite rewarding.   By comparing structures among the reactant and product molecules, one may be able to imagine the kinds of reactive encounters that took place. For example, dimethyl ether decomposes when heated to form ethane, hydrogen, formaldehyde, and methane; it is a simple and informative exercise to invent elementary steps forming these molecules assuming the participation of the radical species H, $CH_3 \cdot$ and $CH_3O \cdot$.

In studying reactions of molecules that have geometrical or optical isomers, the stereochemistry of the product molecules may provide definitive information about the encounter dynamics as well as about the mechanism. We leave detailed consideration of this important topic to its more proper place in organic chemistry, however.

Beware. If mechanistic information is sought through product analysis, it is essential that the reaction time be kept short compared with the time needed to approach equilibrium. Otherwise one may be analyzing the equilibrium mixture and obtaining thermodynamic rather than kinetic information.

Detection of intermediates. Most reaction mechanisms involve other molecules, or molecular fragments, besides those appearing in the chemical equation. Usually, these *intermediate species*, or *intermediates*, are so reactive that detectable concentrations of them do not occur; their transient existence can only be inferred from the kinetic and/or other data. Sometimes, however, in refined experiments, intermediates are found in measurable amounts. Oxygen atoms were observed by mass spectrometry in the reaction $2N_2O \rightarrow 2N_2 + O_2$; CH radicals were detected by absorption spectroscopy in the detonation reaction $\frac{2}{3}C_2H_2 + O_2 \rightarrow 2CO + H_2O$; methyl radicals were found by electron spin resonance of methane samples under bombardment by high energy electrons. If the intermediates of a reaction mechanism can actually be detected, the mechanism is on a far firmer basis than in the usual case in which the identity of the intermediates can only be surmised.

Some free radical or atom intermediates make frequent appearances in reaction mechanisms and have been found to undergo characteristic elementary steps. Therefore, indirect evidence for such steps can be found by comparing features of related reactions; large parts of advanced texts of chemical kinetics are devoted to this topic.

Special indirect chemical tests for the participation of particular kinds of intermediates prove to be useful. For example, triplet state intermediates can be detected by their effect in isomerizing small additions of *cis*-butene to *trans*-butene; aliphatic free radicals in the gas phase react with fresh metal deposits to form metal alkyls, and with toluene to form alkanes and dibenzyl; hydrogen atoms react with iodine to form hydrogen iodide.

Specific reactive intermediates can be intentionally introduced into reactive systems by special techniques. Oxygen atoms can be introduced by irradiating nitrous oxide with ultraviolet (uv) light or by thermal decomposition of ozone; methylene radicals can be introduced by uv irradiation or thermal decomposition of diazomethane; methyl radicals can be introduced by uv irradiation of azomethane; hydrogen atoms can be introduced by uv irradiation of a mixture of mercury vapor and hydrogen.

Isotopes and tracers. The single most important use of isotopes in chemical research is in studying reaction mechanisms. If the special isotope used to prepare reagents for an experiment is not radioactive, the distribution of the isotope in product molecules is determined mass-spectrometrically or by nuclear magnetic resonance (NMR). Frequently used isotopes are deuterium, carbon-13, nitrogen-15, and oxygen-18. Radioactive isotopes used in mechanism studies are called *tracers*; carbon-14 is the best known example. The great utility of tracers is that the measuring instruments used to locate them in the product molecules can be extremely sensitive.

As an example of the use of isotopes, consider the esterification

$$ROH + R'—COOH \rightarrow R'—COOR + H_2O$$

Does the oxygen atom in the water molecule come from the alcohol or from the acid? This question has been answered for many different acids and alcohols by esterifying alcohols whose oxygen-18 content was enriched over the natural abundance level, electrolyzing the water produced, and analyzing the evolved oxygen with a mass spectrometer.

The use of carbon-14 as a tracer is illustrated by a mechanism study of the Claisen rearrangement

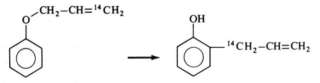

A ^{14}C labeled terminal carbon atom on the ether appears adjacent to the ring in the product phenol, thus suggesting a mechanism involving a cyclic intermediate.

Spectroscopy. The characteristic emission and absorption spectra of molecules often provide direct evidence of the effects of encounters between the emitting or absorbing molecules and other molecules in the system under observation. In brief, the significant aspects are the shapes (widths) of spectral lines, the qualitative appearance of spectra, and the effects of concentration changes upon the line shapes, the line intensities, and the appearance of the spectrum. Spectroscopic information about encounters can refer to properties of individual molecular quantum states, rather than to average properties as does information obtained by the aforementioned methods.

Examples of the use of spectroscopy for obtaining mechanistic information are as follows: proton magnetic resonance observations of H-atom or proton exchange rates in solution, electron spin resonance observations of water-of-hydration exchange rates on paramagnetic ions; microwave line width studies of the effects of encounters on rotational quantum states; ion cyclotron resonance spectroscopic studies of reactions between ions and molecules. These methods are discussed in Chapter 9.

A specialized form of spectroscopy useful for measuring rates of fast reactions is the absorption of ultrasonic power in liquids. By making accurate measurements of the variation with frequency of the absorption coefficient or the sound velocity, chemical reactions with characteristic reaction times near the period of the sound waves can be detected and studied. This kind of experimentation requires different theoretical methods for interpretation of results than the rest of spectroscopy. A brief introduction is given in Chapter 9.

Crossed molecular beams. Since we intend to interpret chemical reactions as net results of large numbers of molecular encounters, it is natural to inquire into the possibilities of obtaining useful information by studying single reactive encounters in the same way that physicists study nuclear reactions, namely, by directing beams of charged particles (neutral molecules) onto target nuclei (other molecules). A molecular beam can be formed by allowing molecules to effuse into a vacuum from a small hole located a short distance from baffles that stop all effusing molecules except those headed in a particular direction (and, in refined experiments, with a certain velocity). If this beam crosses another beam of molecules with which its own molecules can react, then reactive encounters may occur; product molecules will leave the scene of the encounter in directions, and with velocities, determined by conservation of momentum and energy. Movable detectors located a short distance away from the beam intersection can measure the arrival rate of product molecules as a function of angle, and in very refined experiments as a function of product molecule velocity also. The information obtained, expressed in the language of nuclear physics as a reaction cross section, may include the relative distribution of energy of reaction into vibrational, rotational, and translational motions as well as the probability of an encounter leading to reaction.

Unfortunately, this method has only been applied to a limited class of elementary reactions. This is because of experimental complexities arising from the facts that very few encounters take place per second and that the experiment must actually be done not in a void but in a real-life vacuum chamber, containing, typically, about 10^9 molecules/cc background gas. Examples of elementary reactions studied in crossed molecular beams are given in Chapter 9.

SUPPLEMENTARY REFERENCES

Many general chemistry textbooks have introductory discussions of chemical kinetics. Some of them are listed here.

F. Brescia, J. Arents, H. Meislich, and A. Turk, *Fundamentals of Chemistry* (Academic Press, New York, 1966), Chapter 23.

E. L. King, *How Chemical Reactions Occur* (W. A. Benjamin, Inc., New York, 1963).

B. H. Mahan, *University Chemistry* (Addison-Wesley, Reading, Mass., 2nd Ed. 1969), Chapter 9.

C. S. Patterson, H. S. Kuper, and T. R. Nanney, *Principles of Chemistry* (Appleton-Century Crofts, New York, 1967), Chapter 21.

M. J. Sienko and R. A. Plane, *Chemistry: Principles and Properties* (McGraw-Hill, New York, 1966), Chapter 10.

Some scientific papers describing important discoveries about the principles of chemical kinetics have been published as a collection:

M. H. Back and K. J. Laidler, Eds., *Selected Readings in Chemical Kinetics* (Pergamon, London, 1967).

Chapter 2

RATE LAWS

ATTAINMENT OF chemical equilibrium can be viewed in two different ways: either as a consequence of spontaneous processes acting within a thermodynamic system to bring certain state variables to extremum values, or as a consequence of chemical reactions coming to a balance of forward and reverse rates. Let us consider first the thermodynamic point of view.

In a system at constant volume and temperature the formal thermodynamic interpretation is that the Helmholtz free energy (work function) A has its minimum value at equilibrium. Let us consider such a system in which a single chemical reaction occurs and define the *extent of reaction variable* ξ as follows. When the reaction proceeds from left to right as written in the chemical equation to the extent that the number of moles of each reactant consumed is equal to the stoichiometric coefficient of that reactant in the chemical equation, and the number of moles of each product is consequently increased by the corresponding stoichiometric coefficient, then the extent of reaction variable increases by one. For simplicity let us first consider the case in which there is just enough matter in the system that the chemical reaction can proceed only once in the molar amounts indicated by the chemical equation. We can then let the condition $\xi = 0$ indicate that all of the matter is present in the form of reactant molecules. According to our definition the condition $\xi = 1$ will then correspond to all of the matter in the system being in the form of product molecules. Values of ξ between 0 and 1 are linear measures of intermediate compositions. Let us take as a specific

17

example the chemical reaction

$$C_5H_{11}OH + CH_3COOH \rightarrow C_5H_{11}COOCH_3 + H_2O$$

We may let $\xi = 0$ denote that the system contains one mole each of alcohol and acid, $\xi = 1$ denote that the system contains only amyl acetate and water, and $0 < \xi < 1$ denote intermediate compositions.

Values of ξ ranging from 0 to 1 can also be used to describe a system in which one or more of the reactants are in excess and only one of the reactants is present in the stoichiometric amount at $\xi = 0$. Nonreacting species, such as solvents, may also be present in any amounts. There will always be a means of relating ξ to the concentrations of the substances participating in the reaction. In the esterification example given above, we may let the concentration of amyl alcohol at $\xi = 0$ be $[C_5H_{11}OH]_0$ and the concentration of acetic acid at $\xi = 0$ be $[CH_3COOH]_0$. The formulas for intermediate values of ξ may depend on which of the two reactants is in excess at $\xi = 0$. If $[C_5H_{11}OH]_0 \leqslant [CH_3COOH]_0$, then ξ values denoting partial conversion can be calculated from the formula $\xi = [C_5H_{11}COOCH_3]/[C_5H_{11}OH]_0$, where the subscript 0 refers to the amyl alcohol concentration before any esterification has occurred. If amyl alcohol is in excess, then $\xi = [C_5H_{11}COOCH_3]/[CH_3COOH]_0$.

EXERCISE 2-1   Derive computational formulas for $\Delta\xi$ for the arbitrary stoichiometry

$$aA + bB \rightarrow cC$$

assuming that reactant B is present in excess.

$$\xi = \frac{[C]}{[A]_0} \cdot \frac{a}{c}$$

A graph of $A(\xi)$ versus ξ for our simple example is shown in Figure 2-1. At the composition ξ_{eq} the system does not change with time; at any other value of ξ, the system composition will change with time in the direction *toward ξ_{eq}. If the A(ξ) graph and the value of ξ are known, the direction of change of ξ with time is known. What we seek now is knowledge about the rate at which ξ approaches ξ_{eq}; in the language of calculus, we need to study the time derivative of ξ, $d\xi/dt$.*

In many systems the rate of approach to equilibrium is so slow that, for all practical purposes, the time derivative of ξ appears to be zero. For example, the free energies of formation of substances such as O_3, C_2H_2, and NH_4NO_3 are positive; according to the argument of the preceding paragraph, these compounds must revert to the elements, reducing $A(\xi)$ to zero, or to more stable compounds, reducing $A(\xi)$ to a negative value. The reaction rates at room temperature, however, are actually so slow that pure samples of these compounds may be stored indefinitely with no trace of reaction; that is, to the best of our ability to analyze the samples, $d\xi/dt = 0$. In this way we see that the chemical composition of matter is governed not only by thermo-dynamics, but also by the rates of those reactions that are thermodynamically possible. A corollary of this idea is that the products formed in chemical reactions between a given set of reactants may be formed in quantities governed more by the relative rates of the various reactions that can occur than by the thermodynamic properties of the possible products. This is a striking characteristic of living organisms. As a specific example, thermo-dynamic consideration of any system comprised of sugar plus an excess of oxygen would lead to the conclusion that chemical reaction necessarily leads to the formation of carbon dioxide and water as the only products. Metabol-ism of sugar in the human body, however, actually leads to production of many other compounds besides carbon dioxide and water. Likewise, com-bustion of gasoline with excess air produces only carbon dioxide and water according to thermodynamics, but in fact produces small quantities of carbon monoxide and low molecular weight hydrocarbons as well. In cases such as these we say that the over-all reactions are *kinetically controlled* to give product distributions other than the one which thermodynamics alone would lead us to expect. One might say that thermodynamics by itself characterizes *possible* chemical reactions, *not real* chemical reactions.

There are many physical quantities that can be written as single-valued functions of ξ or, conversely, that could be used as experimental means of measuring ξ. Some examples would be density, temperature, absorption spectrum, or index of refraction. Reactant or product concentrations, how-ever, turn up in reaction rate problems more frequently than any other

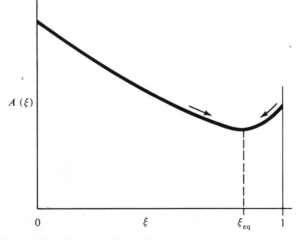

FIGURE 2-1 *Graph of Helmholtz free energy A versus extent of reaction variable ξ for a chemical reaction taking place at constant temperature and volume. The graph of Gibbs free energy versus ξ for a reaction taking place at constant temperature and pressure has the same qualitative features as this one.*

measures of ξ, for a variety of practical reasons. Concentrations are also convenient variables to use in theories of elementary reaction rates and for interpreting the dependence of conversion rates upon reaction conditions. In this book we generally use moles per liter concentration units and denote concentrations by square brackets [ ]. Unless stated otherwise, we consider time derivatives of molar concentrations, $d[\ ]/dt$, as our measures of $d\xi/dt$. In theoretical derivations it will be expedient to use molecules per cubic centimeter concentration units, denoted ^1N, and to express rates as $d\,^1N/dt$.

Since we are dealing in this chapter mostly with general principles rather than with applications to particular chemical reactions, we denote elements and compounds by general symbols, such as A, B, and C, rather than by the usual chemical symbols for particular elements and compounds. In Chapter 3 we introduce specific examples and replace the general concentrations, such as [A], appearing in this chapter with concentrations of specific substances, such as $[Br_2]$ or $[C_5H_{11}COOCH_3]$.

2-1 THE PRINCIPLE OF MASS ACTION AND THE RATE LAWS OF ELEMENTARY REACTIONS

Suppose that the chemical reaction taking place is $2A + 3B \rightarrow C$ and that the concentrations at $t = 0$ are $[A]_0$, $[B]_0$, and $[C]_0$. Values of $\Delta\xi$ as reaction proceeds are given by any of the expressions $-\frac{1}{2}V([A] - [A]_0)$, $-\frac{1}{3}V([B] - [B]_0)$, or $V([C] - [C]_0)$. The derivatives $d[A]/dt$, $d[B]/dt$, and $d[C]/dt$ are all proportional to $d\xi/dt$, but the constant of proportionality is different in each case. If we wish to replace $d\xi/dt$ with a more practical $d[\ ]/dt$, then we must decide which concentration is meant. To avoid trouble, the following convention is adopted. The *conversion rate*, in concentration per time units, is defined as *the time rate of change of* any *reactant or product concentration divided by the stoichiometric coefficient of that reactant or product, with the coefficient being taken as positive for products and negative for reactants*; in this way conversion rates are equal whichever concentration is taken. Symbolically, if v_i and A_i are the stoichiometric coefficient and the chemical symbol of substance i, respectively, the stoichiometric equation for the reaction is

$$\boxed{\sum v_i A_i = 0}$$

and the conversion rate definition is

$$\boxed{R \equiv \frac{1}{v_i}\frac{d[A_i]}{dt}}$$

for any i.

Now let us turn to the second aspect of equilibrium: the balance of forward and reverse reaction rates. In the terms just introduced, this provides a succinct definition of chemical equilibrium, namely, $R = 0$. We can expand the notion of conversion rate to separate, conceptually, the forward and reverse reaction rates. This is visualized most easily in terms of the elementary reactions: at equilibrium the net effect $R = 0$ is attained because for each elementary reaction the rate of making reactant molecules into product molecules is equal to the reverse rate of making product molecules into reactant molecules.

For the special case in which the stoichiometric equation and the mechanism are one and the same thing, this aspect of chemical equilibrium leads immediately to the *Law of Mass Action*. A function of temperature called the *equilibrium constant*, defined as

$$K_{eq}(T) = \prod [A_i]_{eq}^{\nu_i}$$

where the product is taken over the concentrations of all species appearing in the stoichiometric equation, is considered to be a ratio of rate constants $k_f(T)/k_r(T)$, and the equal forward and reverse rates at equilibrium are written

$$R_{f, eq} = k_f(T) \prod_{reactants} [A_i]_{eq}^{-\nu_i}$$

and

$$R_{r, eq} = k_r(T) \prod_{products} [A_i]_{eq}^{\nu_i}$$

The net rate R_{net} being zero at equilibrium

$$R_{net, eq} = R_{f, eq} - R_{r, eq} = 0$$

leads to

$$k_f(T) \prod_{reactants} [A_i]_{eq}^{-\nu_i} = k_r(T) \prod_{products} [A_i]_{eq}^{\nu_i}$$

or

$$\frac{k_f(T)}{k_r(T)} = \prod_{\substack{reactants \\ and \\ products}} [A_i]_{eq}^{\nu_i} = K_{eq}(T)$$

We write K_{eq}, k_f and k_r as functions of the temperature T to emphasize that K_{eq} changes with temperature (according to the Gibbs–Helmholtz equation) and that at least one of k_f, k_r must likewise vary with T.

For *elementary reactions*, we can extend the notion of forward and reverse rates to *nonequilibrium* situations and write

$$R_f = k_f(T) \prod_{\text{reactants}} [A_i]^{-\nu_i}$$

and   Elementary reactions only; equilibrium or no equilibrium.

$$R_r = k_r(T) \prod_{\text{products}} [A_i]^{\nu_i}$$

Until equilibrium has been established, the nonzero net rate $R_{net} = R_f - R_r$ describes the observed rate of conversion of reactants into products. These equations comprise the *Principle of Mass Action*, which expresses in symbols the idea introduced in Chapter 1 that *for any elementary reaction the number of reactant molecules converted per second to product molecules is proportional to the concentration of each substance participating in that elementary reaction*, since the encounter frequency will be proportional to each such concentration.

We have deduced from the dynamic aspect of a particular type of chemical equilibrium, where stoichiometric equation and mechanism are identical, a useful functional form relating conversion rates to reactant concentrations. The R_f function contains the concentrations of the molecules appearing on the left-hand side of the reaction equation, and the reverse reaction rate R_r is a function of the concentrations of the molecules appearing on the right-hand side of the reaction equation. In either case, the rate function really depends on the relevant *reactant molecule concentration only*, since the products of the forward reaction are the reactants of the reverse reaction. Furthermore, the functional form is dictated by these considerations to be a product of concentrations raised to integral powers; as indicated in Chapter 1, the possible powers are 1 (the usual case), 2 (infrequent), or 3 (only a handful of known cases). To the extent that it is true that the stoichiometric equation and the mechanism are identical, the R functions are also necessarily correct for the two rates, and they have acquired the name of *rate law* on this account. *The term* rate law *should not be confused with actual laws of nature; consider it to be synonymous with "rate-of-conversion function."*

The concept, and to some extent the functional forms as well, of rate laws may now be extended to describe *any* conversion rates at all, in particular to describe the dependence of experimental conversion rates upon concentrations. We shall now and hereafter differentiate between experimental rate laws and rate laws predicted on the basis of an assumed reaction mechanism.

Experimental rate laws will have concentration exponents written as *decimal numbers.* For example, the experimental rate law for the reaction

$$H_2 + D_2 \rightarrow 2\,HD$$

in the presence of argon was found to be

$$R \equiv \tfrac{1}{2}\frac{d[HD]}{dt} = k(T)\,[H_2]^{0.38}\,[D_2]^{0.66}\,[Ar]^{0.98}$$

where $k(T) = 10^{9.54}\,T^{\frac{1}{2}}\exp(-42{,}260/1.987\,T)$ liter/mole-sec, 0.38, 0.66, and 0.98 are experimental quantities. *Rate laws predicted on the basis of an assumed mechanism* will have concentration exponents written as *integers or fractions.* For example, one proposed mechanism (to be discussed in Chapter 3) for the chemical reaction

$$CO + Cl_2 \rightarrow COCl_2$$

leads to the predicted rate law

$$R = \frac{d[COCl_2]}{dt} = k_3(T)K_1(T)^{\frac{1}{2}}K_2(T)\,[CO]^{(1)}[Cl_2]^{\frac{3}{2}}$$

The subscripts refer to three elementary steps assumed to comprise the mechanism of the reaction forming $COCl_2$; the rate constant $k_3(T)$ and the equilibrium constants $K_1(T)$ and $K_2(T)$ all appear in the rate law, rather than a single $k(T)$; the exponents (1) and $\frac{3}{2}$ were predicted by analyzing the mechanism with methods we discuss later.

A few words about notation are appropriate at this point. The functions $k(T)$ appearing in the equations of this section are numerically equal to the elementary reaction rates that would be measured if each of the concentrations in the rate laws were 1 molar, or more generally if the indicated products of concentrations happened to be unity. For this reason these functions are occasionally called *specific rate constants.* Two other names are *rate coefficient* and *velocity constant.* The term *rate constant* is by far the most common one, however, and we accept the general usage here. *We must remember that rate constants are in fact functions: they always vary with temperature, usually rapidly, and sometimes with other variables as well.* As a reminder of this, the dependence upon temperature—the most important variable—will usually be written out explicitly in this book as $k(T)$, even though k alone is customary practice elsewhere. The customary misnomer "rate constant" can be accepted in much the same sense that "weight" is customarily accepted when mass is actually meant.

The sum of the exponents appearing in a rate law for a single elementary reaction gives the *molecularity* of that elementary reaction; the sum is one for a *unimolecular* reaction, two for a *bimolecular* reaction, and three for a *termolecular* reaction. The analogous sum for an experimental rate law is called the *order* of the reaction. If the decimal sum happens to come out, within experimental error, to have an integral or a simple fractional value, the order may be so stated; for example a *first order* reaction has the exponent sum 1.0, a *three-halves order* reaction 1.5, and so on. It is also common to distinguish between the sum and its terms as *over-all* order and order *with respect to* a reactant. In the $H_2 + D_2$ example given above, the reaction is *second order overall, first order in argon*, and has orders 0.38 and 0.66 with respect to hydrogen and deuterium, respectively. It is frequently the case, however, that the functional form of the experimental rate law does not lend itself to description by order; moreover, the order may change with reactant concentrations owing to changes in the mechanism. Rate laws that are predicted from assumed mechanisms may or may not lend themselves to description by order.

EXERCISE 2-2   A predicted rate law for the reaction $H_2 + Br_2 \rightarrow 2HBr$ is

$$R \equiv \tfrac{1}{2}\frac{d[\text{HBr}]}{dt} = \frac{k_a(T)\,[\text{H}_2][\text{Br}_2]^{\frac{1}{2}}}{1 + k_b(T)\,[\text{HBr}]/[\text{Br}_2]}$$

The complicated dependence of R on $[\text{Br}_2]$ becomes quite simple for two limiting cases given by extreme values for the ratio $[\text{HBr}]/[\text{Br}_2]$. What is the approximate over-all order predicted for the early stage of reaction, where $[\text{HBr}] \ll [\text{Br}_2]$? What order in bromine is predicted if the reverse inequality $[\text{HBr}] \gg [\text{Br}_2]$ holds?

   Experimental measurement of reaction orders with respect to all concentrations is the usual starting point for finding a reaction mechanism. In Chapter 3 we consider various means of carrying out these measurements.

   We conclude these introductory considerations of rate laws with some words of caution.

   It is important to realize that experimental rate laws only describe conversion rates for the period during which the conversion rates are actually measured. It is possible and in general likely that the conversion rates at other times follow quite different rate laws. At the start of any reaction, there are deviations caused by such factors as the turbulence developed in mixing the reagents, establishing thermal equilibrium, or first establishing chemical equilibrium in "fast" reactions, such as those reactions involved in buffering solutions to constant pH, which then remain fully equilibrated while the progress of the reaction of interest is being measured. After a large fraction of reaction has occurred, such factors as slower side reactions and reverse reactions may become important and invalidate the experimental rate law as a description of the conversion rate at long reaction times. Furthermore, considerations such as these apply to predicted rate laws as well, because mechanisms and the rate laws derived from them are at their very best only approximations. Anyone doing chemical kinetics experiments who neglects these facts will sooner or later find himself in trouble. Do not lose sight of the inherent limitations of experimental and predicted rate laws.

We next consider the evolution in time of reacting systems whose experimental rate laws have simple (near-integral) orders. Our goal is to obtain explicit functions of time, which may be called *integrated rate laws*, from the differential equations of the simple rate laws. These differential equations will all be solved using the so-called separation of variables method. More refined mathematical techniques, such as the operator or Laplace transform methods, are necessary to integrate the differential equations of complicated rate laws.

2-2 FIRST ORDER REACTIONS

The rate law for a reaction that is first order in substance A is

$$R \equiv \frac{1}{v_A} \frac{d[A]}{dt} = k(T) [A]^{1.0}$$

If A is one of the reactants appearing in the stoichiometric equation, or the only reactant, v_A is negative; usually, $v_A = -1$. If it is not -1, the following equations must be modified to include v_A. In the $v_A = -1$ case we have

$$R \equiv \frac{-d[A]}{dt} = k(T) [A]^{1.0}$$

We integrate this rate law from $[A]_0$, the concentration of A at the time reaction starts, $t = 0$, to the variable upper limit $[A]$, a function of t, on the assumption that T is constant:

$$-\int_{[A]_0}^{[A]} \frac{1}{[A]} d[A] = \int_0^t k(T) \, dt$$

$$\ln \frac{[A]_0}{[A]} = k(T)t$$

$$[A] = [A]_0 \exp[-k(T)t]$$

An exponential decay of [A] in t thus characterizes first order reactions in which v_A is negative. For experimental determination of $k(T)$, the integrated rate law is conveniently written

$$\log_{10}[A] = \log_{10}[A]_0 - \frac{k(T)t}{2.303}$$

A semilogarithmic plot of [A] versus t is a straight line with slope $-k(T)/2.303$ and intercept $[A]_0$. Such a plot is shown in Figure 2-2.

An interesting special case of a first order reaction is nuclear fission, such as

$$n + {}^{235}U \rightarrow {}^{97}Zr + {}^{137}Te + 2n$$

where n is a neutron and $[{}^{235}U]$ is essentially constant. For this reaction v_n is $+1$. The integrated rate law shows exponential *growth* of $[n]$, that is, an explosion.

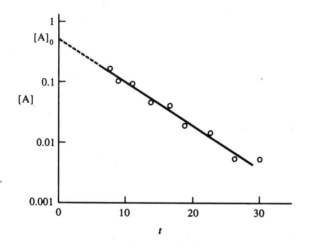

FIGURE 2-2 *Schematic graph of concentration versus time for a substance that is disappearing in an "irreversible" first order reaction. In the case shown here the concentration decrease was measured over about one and one-half decades. For an adequate measurement of the first order rate constant the concentration range studied should be at least this large. Note that it is not necessary to know an absolute concentration in order to evaluate a first order rate constant: relative concentrations as functions of time are sufficient. In practice this may mean that the inital value of the decaying concentration $[A]_0$ need not be known. If of interest, however, it can be found from a concentration measurement and a rate constant measurement made at a later time. This is the basis of age measurements in geology, which utilize the first order decay rates of naturally occurring radioactive isotopes.*

2-3 SECOND ORDER REACTIONS

We must consider three cases of second order reactions. The first case is a reaction second order in the concentration of a single reactant; the second case is a reaction second order overall and first order in the concentrations of each of two reactants whose initial concentrations are in the same ratio as their stoichiometric coefficients; the third case is the same as the second except that the initial concentrations are not in this ratio.

The rate law for the first case is

$$R \equiv \frac{1}{\nu_A} \frac{d[A]}{dt} = k(T) \, [A]^{2.0}$$

and ν_A is negative for all known examples (usually $\nu_A = -1$ or -2). The integration is straightforward

$$\int_{[A]_0}^{[A]} [A]^{-2.0} \, d[A] = \int_0^t \nu_A \, k(T) \, dt$$

$$\frac{1}{[A]} = \frac{1}{[A]_0} - \nu_A \, k(T)t$$

A plot of inverse concentration versus time yields a graph with slope $-\nu_A \, k(T)$ and intercept $1/[A]_0$. A schematic graph of this type is shown in Figure 2-3.

The rate law for the second case

$$R \equiv \frac{1}{\nu_A} \frac{d[A]}{dt} = k(T) \, [A]^{1.0} [B]^{1.0}$$

can be converted to that for the first case by noting that the concentration of B is related to the concentration of A by the relation $[B] = \nu_B/\nu_A[A]$ throughout the course of the reaction. The result is

$$\frac{1}{[A]} = \frac{1}{[A]_0} - \nu_B \, k(T)t$$

The rate law for the third case

$$R = \frac{1}{\nu_A} \frac{d[A]}{dt} = k(T) \, [A]^{1.0} [B]^{1.0}$$

must be altered to make the right-hand side a function of only variable one before integration can be carried out. Define $x \equiv$ number of moles A reacted at a given time; then $[A] = [A]_0 - x$ and $[B] = [B]_0 - (v_B/v_A)x$.

$$\frac{1}{v_A}\frac{d([A]_0 - x)}{dt} = \frac{-1}{v_A}\frac{dx}{dt} = k(T)\,([A]_0 - x)\left([B]_0 - \frac{v_B}{v_A}x\right)$$

The integrated rate law is complicated:

$$\ln\left[\frac{v_B\,[A]_0 - v_A\,[B]_0}{[A]} - v_B\right] = \ln\left[\frac{-v_A\,[B]_0}{[A]_0}\right] - \left(v_A\,[B]_0 - v_B\,[A]_0\right)k(T)t$$

EXERCISE 2-3 Work out the integrations in detail for each case of second order reaction. You will need to know the mathematical formulas for a partial fraction expansion. They are as follows. For a function $f(x)$ of the form

$$f(x) = \prod_{i=1}^{N} (a_i x - b_i)^{-1}$$

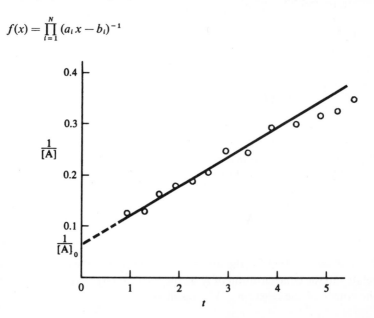

FIGURE 2-3 *Schematic graph of inverse concentration versus time for a substance A that is disappearing by a reaction that is second order in A. In the example shown the data points for long reaction times do not fit on the straight line. This indicates that the mechanism valid for describing this chemical reaction at short reaction times is not valid for long times. For determining the value of a second order rate constant it is necessary to measure absolute concentrations rather than relative concentrations as in the case of a first order reaction. Complicated functions may have to be plotted if one wishes to graph second order decay plots for reactions that are first order in two substances and second order overall.*

where there are no recurring factors, that is, no repeating (a_i, b_i), the partial fraction expansion is

$$f(x) = \sum_{i=1}^{N} c_i[a_i x - b_i]^{-1}$$

where

$$c_i = \prod_{\substack{k \neq i \\ k=1}}^{N} \left(\frac{a_k}{a_i} b_i - b_k\right)^{-1}$$

2-4 THIRD ORDER REACTIONS

Rate laws for third order reactions can be integrated readily. Complicated expressions can arise, however. For the simple case

$$R \equiv \frac{1}{v_A} \frac{d[A]}{dt} = k(T) \, [A]^{1.0} \, [B]^{1.0} \, [C]^{1.0}$$

with $v_A = v_B = v_C = -1$ and $[A]_0 = [B]_0 = [C]_0$

$$\frac{1}{[A]^2} = \frac{1}{[A]_0^{\,2}} + 2k(T)t$$

A plot of inverse concentration squared versus time yields a graph with slope $2k(T)$.

EXERCISE 2-4 Work out the integrated rate law for a reaction $A \rightleftharpoons B$ taking explicit account of the reverse reaction, for the case that the forward reaction is first order in A and the reverse reaction is first order in B, assuming $[B]_0 = 0$. Sketch a graph of [A] and [B] versus time for the function you obtain.

EXERCISE 2-5 Work out the integrated rate law for concurrent first order reactions (in A) yielding two different products, assuming that reverse reactions may be neglected and that the product concentrations are zero at the start of the experiment. Use the outline mechanism

$$A \xrightarrow{k_1(T)} B$$

$$A \xrightarrow{k_2(T)} C$$

Find [B] (t) and [C] (t) as well as [A] (t). Draw a graph of the functions you obtain. Also, sketch graphs of the free energy functions $A_1(\xi_1)$ and $A_2(\xi_2)$ versus ξ_1 and ξ_2 that would justify neglecting the reverse reactions.

2-5   *PREDICTION OF RATE LAWS FOR ASSUMED MECHANISMS*

The basic principle underlying the derivation of the rate law for a mechanism is the *Principle of Mass Action*, discussed in Section 2-1: the rate of any elementary reaction is proportional to the frequency of encounters indicated by the chemical equation of the elementary reaction, hence to the indicated product of concentrations.   In order to facilitate use of this principle in deriving rate laws we have to add to it three subsidiary notions.

First, it is sometimes possible to *neglect back reactions*.   This amounts to an *ad hoc* postulate that, for practical purposes, a given elementary reaction proceeds only in one direction.   This will clearly be a satisfactory assumption in the early stages of any reaction, since only low concentrations of the product molecules will be available to serve as reactants for back reactions.   It will also be true for later stages if the elementary reaction equilibrium constant K_{eq} is large, since in that case the large forward rate constant $k_f = K_{eq} \times k_r$ will keep the forward rate much larger than the reverse rate until the product concentrations are much larger than the reactant concentrations.

The postulate that a given elementary reaction can be considered to proceed in only one direction is often expressed as an assumption that the reaction is "*irreversible*."   We shall occasionally set "irreversible" in this sense in quotation marks to prevent possible confusion with the entirely different meaning of the term irreversible in thermodynamics.

The second idea is the *rate limiting step*.   A mechanism may contain a sequence of elementary reactions one of which has a much slower rate than the others.   The conversion rate can be correctly calculated on the assumption that this single step controls it, and that all other steps are either faster, completely "irreversible" ones or faster, completely "reversible," that is, equilibrated ones.

The third idea is called the *steady state approximation*.   In a sequential mechanism, such as

$$A \rightarrow B$$

$$B + A \rightarrow C$$

it may happen that one or more of the intermediate species has equal rates of production and destruction, after a period of initial concentration increase that is short compared to the over-all conversion time.   Setting these two rates equal to one another usually permits simplification of the rate laws by removing the (unknown) intermediate concentrations from the equations. This assumption that intermediate concentrations are at a steady state value can sometimes be subjected to experimental test.

Before proceeding with illustrations of these ideas, it should be made clear that using mathematical approximations for obtaining rate laws from an assumed mechanism is not really a prerequisite for comparing the mechanism with rate data. The set of differential equations corresponding to any mechanism merely can be solved directly with an analog or a digital computer. A computer can predict all the concentration versus time graphs, given the mechanism, the starting concentrations, and the $k(T)$ values. Moreover, a computer program can easily take into account volume and temperature changes, flow effects, and so on. The drawback of using computers to determine the predictions of mechanisms is that too much information must be given to the computer and that too much is returned. Extracting the *significant* information from computer output is likely to be more difficult than analyzing the differential equations in the first place. When the differential equations cannot be understood, however, computed solutions can be quite helpful, or indeed the only path for progress.

To illustrate the reasoning used in proceeding from an assumed mechanism to a predicted rate law, let us consider two examples. Acetaldehyde in aqueous solution exists in hydrated and nonhydrated form. The mechanism of hydration, which was studied by NMR line width experiments as well as by traditional methods, is assumed to be

$$CH_3CHO + H_3O^+ \underset{k_{BA}}{\overset{k_{AB}}{\rightleftharpoons}} CH_3CHOH^+ + H_2O$$
$$\quad\text{(A)} \qquad\qquad\qquad\qquad\quad \text{(B)}$$

$$CH_3CHOH^+ + H_2O \underset{k_{CB}}{\overset{k_{BC}}{\rightleftharpoons}} CH_3CH(OH)OH_2^+$$
$$\quad\text{(B)} \qquad\qquad\qquad\qquad\quad \text{(C)}$$

$$CH_3CH(OH)OH_2^+ + H_2O \underset{k_{DC}}{\overset{k_{CD}}{\rightleftharpoons}} CH_3CH(OH)_2 + H_3O^+$$
$$\quad\text{(C)} \qquad\qquad\qquad\qquad\quad\quad \text{(D)}$$

The first manipulation is to define simple symbols for the seven substances and the six rate constants. It is convenient to name the four aldehyde species A, B, C, and D as indicated. The necessary equations for $d[A]/dt$, $d[B]/dt$, $d[C]/dt$, and $d[D]/dt$ can be written as if all the steps were unimolecular if we assume that the proton and water concentrations are constants that can be combined with the corresponding rate constants. Thus, the rate law for A

$$\frac{d[A]}{dt} = -k_{AB}(T)\,[A][H_3O^+] + k_{BA}(T)\,[B][H_2O]$$

becomes

$$\frac{d[A]}{dt} = -k'_{AB}[A] + {}'_{BA}[B]$$

where

$$k'_{AB} = k_{AB}(T)\,[H_3O^+]$$

and

$$k'_{BA} = k_{BA}(T)\,[H_2O]$$

The rate constants are subscripted to indicate the elementary steps, primed as a reminder of the inclusion of hydronium ion and water concentrations in each one, and written without explicit mention of their temperature dependence.   We can now discuss parts of the mechanism, such as $C \rightleftarrows D + H^+$, at constant pH without writing down the H^+, H_3O^+, and H_2O participation each time.

Next we simplify the problem by utilizing knowledge of chemistry.   Since protonated acetaldehyde hydrate is a strong acid, the equilibrium $C \rightleftarrows D$ is established very rapidly compared with the rest of the elementary reactions. The rate of the hydration reaction $A \rightarrow D$ is, therefore, determined entirely by the rate of $A \rightarrow C$ conversion.   Thus we can reduce the mechanism either to $A \rightleftarrows B \rightleftarrows C$ or to $A \rightleftarrows B \rightleftarrows D$ and obtain the same rate predictions.   Let us choose the latter since the equilibrium constant for $C \rightleftarrows D$ ensures that $[D] \gg [C]$ will be valid for the pH range used in the experiments.

Acetaldehyde is a very weak base.   This means that protonated acetaldehyde B will have an extremely low concentration compared to acetaldehyde; in the mechanism, B ions can be regarded as having only brief transient existence in $A \rightarrow B \rightarrow C$, $A \rightarrow B \rightarrow A$, $C \rightarrow B \rightarrow C$ and $C \rightarrow B \rightarrow A$ transformations.   The NMR experiments carried out to study this mechanism measure, independently, the over-all rates of the forward and reverse conversions $A + H_2O \rightarrow D$ and $D \rightarrow A + H_2O$.   For the forward conversions, the measured quantity is k_{AD} in the experimental rate law

$$R_{AD} = k_{AD}[A]^{1.0}\,[H_2O]^{1.0}$$

How is the experimental k_{AD} related to the rate constants for the elementary steps?   The intermediate B ions just formed from A can either revert to A by the $B \rightarrow A$ reaction or proceed to D by the $B \rightarrow D$ reaction.   The fraction of B molecules converted to D is equal to the ratio of the rate of $B \rightarrow D$ to the sum of the rates $B \to D$ and $B \to A$, or

$$\text{fraction of B going to D} = \frac{k'_{BD}[B]}{k'_{BA}[B] + k'_{BD}[B]} = \frac{k_{BD}}{k_{BA} + k_{BD}}$$

Combining this ratio with the rate of formation of B from A, $k'_{AB}[A]$, gives the rate of the $A \to D$ forward conversion.

$$R_{AD} = \frac{k'_{AB}[A]k_{BD}}{k_{BA} + k_{BD}}$$

Comparing our R_{AD} equations shows that

$$k_{AD}[H_2O] = \frac{k'_{AB}k_{BD}}{k_{BA} + k_{BD}}$$

$$k_{AD} = \frac{k_{AB}k_{BD}}{k_{BA} + k_{BD}} \frac{[H_3O^+]}{[H_2O]}$$

The assumed mechanism thus provides a connection between the rate constant k_{AD} in the experimental rate law and elementary step rate constants k_{AB}, k_{BA} and k_{BD}. Further interpretation, now well beyond our scope, requires additional knowledge of the chemistry of reactions in aqueous solutions.

Our reasoning will change if the mechanism contains steps assumed to be "irreversible," that is, assumed to proceed in only one direction for the conditions of experimental interest. The most common such mechanism is quite similar to the reversible one just treated, and may be written schematically

$$A \underset{k_{BA}}{\overset{k_{AB}}{\rightleftharpoons}} B \overset{k_{BC}}{\longrightarrow} C$$

Since we have several types of specific examples of this mechanism to consider later, we consider here only the formal reasoning used to find a predicted rate law from the mechanism.

Once again B is an intermediate species with only transient existence and low concentration. If the intermediate B can undergo an irreversible step, such that the rate of formation of B from C is zero, its concentration may be assumed to have a *steady state* value in which the rate of destruction of B ($B \to A$ and $B \to C$) is equal to its rate of formation ($A \to B$). This will be true if, for example, the characteristic times (for example the half-times in

Chapter 3), for the opposing reactions $A \to B$ and $B \to A$ are small in comparison with both the characteristic time for the over-all conversion $A \to C$ *and* the time scale of the experiments. The steady state assumption is, verbally, that the rate of formation of B is equal to its rate of destruction, or, operationally,

$$R_{formation} \text{ of } B = R_{destruction} \text{ of } B$$

$$k_{AB}[A] = k_{BA}[B] + k_{BC}[B]$$

Solving for the steady state concentration of B, we find that

$$[B] = \frac{k_{AB}[A]}{k_{BA} + k_{BC}}$$

The conversion rate, equal to the $B \to C$ rate, is

$$R \equiv \frac{d[C]}{dt} = \frac{k_{BC} k_{AB}[A]}{k_{BA} + k_{BC}}$$

If the experimental rate law were found to be $R = k_{AC}[A]^{1.0}$, then the experimental k_{AC} would be related to the elementary reaction rate constants by

$$k_{AC} = \frac{k_{BC} k_{AB}}{k_{BA} + k_{BC}}$$

The elementary reactions actually may be bimolecular or termolecular, in which case the rate constants written here have incorporated concentrations, as in the equilibrium case previously considered.

In subsequent chapters we shall derive rate laws for many mechanisms. Now it is only important to understand the logical steps involved. Later we shall consider experiments that provide rate laws and interpretations that associate chemistry and physics with elementary reaction rate constants.

2-6 *TEMPERATURE DEPENDENCE OF RATE CONSTANTS*

Reaction rates change with temperature for two reasons: encounter frequencies rise with temperature, and high energy encounters become more probable as the temperature increases. The dependence of encounter frequency on temperature at constant density is a slowly varying function, proportional approximately to $\sqrt{T}$. The probability of high energy encounters varies rapidly with temperature, approximately as $\exp(-\text{constant}/T)$.

We noted earlier that the forward and reverse rate constants of a reaction whose mechanism and stoichiometric equation are identical are related to the equilibrium constant by the equation $K_{eq} = k_f/k_r$, with K_{eq} expressed in concentration units.   If the reactants and products happen to be ideal gases, $K_{eq} = K_p(RT)^{-\Sigma v_i}$ and the temperature dependence of K_{eq} can be obtained from the Gibbs–Helmholtz equation

$$\frac{d \ln K_p}{dT} = \frac{\Delta H}{RT^2}$$

or

$$\frac{d \ln K_{eq}}{dT} = \frac{\Delta H - \sum v_i(RT)}{RT^2} = \frac{\Delta E}{RT^2}$$

If the temperature dependence of ΔE is neglected, an approximate integration from $1/T = 0$ to $1/T$ can be carried out that yields

$$K_{eq}(T) = K_{eq}(1/T = 0)\exp(-\Delta E/RT)$$

The ratio k_f/k_r thus has, approximately, an exponential dependence on inverse temperature.   This suggests that both k_f and k_r may have, approximately, exponential dependence on inverse temperature, or that one of them has the same temperature dependence as the equilibrium constant and the other one is independent of temperature.

EXERCISE 2-6   Complete all steps of the argument presented in the preceding paragraph.

From thermodynamics we are thus led to suspect that ideal gas reaction rate constants may show exponential inverse temperature dependence.   This is confirmed experimentally, not only for gas reactions, but quite generally. In fact, it was first proposed, by Arrhenius in 1889, on the basis of experimental rate measurements rather than thermodynamic intuitions.   The expression stemming from Arrhenius' research is

$$k(T) = A \exp(-E_A/RT)$$

Here, all of the temperature dependence of $k(T)$ is placed in the exponential. We shall find later that there are good theoretical reasons for writing A as a function of temperature also; however, $\exp(-E_A/RT)$ varies so strongly with temperature for $E_A \gg RT$ (the usual case) that experimental data usually do not allow the temperature dependence in $A(T)$ to be evaluated.

(*Note:* We shall use the symbol A henceforth only to denote the quantity appearing in this equation; there should be no reason for confusing it with the Helmholtz free energy for which we have heretofore used the same symbol.)

A schematic Arrhenius graph is shown in Figure 2-4.

The two parameters A and E_A of the Arrhenius expression for $k(T)$ are called "*A factor*" and "*activation energy.*" The A factor is related to the encounter frequency and the encounter geometry, while the activation energy is related to the strength of the chemical bonds that are broken and formed. Interpretation of A and E_A values is the central problem of the theories of reactive encounters we study in later chapters. These theoretical interpretations will allow us to use experimental A and E_A values for obtaining insights into reaction mechanisms beyond those made available from experimental reaction orders.

The function $\exp(-E_A/RT)$ is reminiscent of the Boltzmann distribution law of statistical thermodynamics. The similarity is not coincidental, but

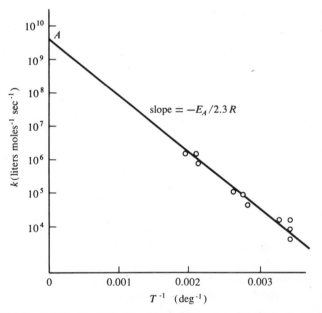

FIGURE 2-4 *Schematic Arrhenius graph for a second order rate constant. Note that a long extrapolation is required to evaluate A even though the experimental data points shown cover the wide temperature range from room temperature up to 500°K (223°C). One consequence of the long extrapolation is that two investigators whose rate measurements yield rate constants that are in quite good agreement with one another may find A and E_A values that differ greatly. Another consequence is that small systematic errors caused by mechanism changes with temperature may greatly distort the A and E_A values even though the rate constant itself is nearly unaffected.*

fundamental to the theories of reaction rates. We shall find that the way in which $\exp(-E_A/RT)$ enters into the $k(T)$ function is the distinguishing characteristic of the different theories.

2-7 SUMMARY

Let us now summarize what has been said about rate laws and reaction mechanisms. *Rate laws* have a twofold aspect. On the one hand they are directly derived from experimental data and serve as factual summaries both for practical purposes and for interpretative use. On the other hand, our confidence that macroscopic changes manifest molecular events tells us that these same rate laws must be conceptually predictable from physical laws describing molecular encounters. *Mechanisms* are purely hypothetical constructs. They serve as bases for discourse on the nature of reactive molecular encounters and for predictions, including rate laws, that can be compared with experiments.

Experimental *rate laws* written in product form have *orders* that are *empirical* and *nonintegral*, though they may in some cases be near-integral within experimental accuracy; the elementary reactions comprising reaction mechanisms have *molecularities* that are *integers*. We considered some simple rate laws and integrated them to obtain equations for the evolution in time of the reacting systems; we anticipate that experimental study of reaction rates will yield data that fit such equations, and thence enable us to determine reaction orders. Whereas reaction orders only provide information about reaction mechanisms, the temperature dependence of rate constants can be taken further, to provide information about the nature of individual reactive encounters themselves.

EXERCISES

2-7 Ozone decomposes when heated to form oxygen. A proposed mechanism is

$$O_3 \rightarrow O_2 + O$$

$$O + O_3 \rightarrow 2O_2$$

Assuming that the first step is irreversible, use the steady state approximation on O atoms to derive the rate law and the steady state value of $[O]$ for this reaction. Sketch graphs of $[O_3]$ and $[O_2]$ versus time, assuming that the decomposition proceeds to completion.

2-8 Hydrogen iodide can be synthesized from the elements by the chemical reaction

$$H_2 + I_2 \rightarrow 2HI$$

One proposed mechanism for the reaction is

$$I_2 \rightarrow 2I \tag{1}$$

$$H_2 + 2I \rightarrow 2HI \tag{2}$$

Derive rate laws assuming (a) that reaction (1) is in equilibrium and (b) that a steady state approximation applies on I atoms and $k_2[H_2] \gg k_{-1}$. Describe how you could determine experimentally which of the two assumptions is correct.

2-9   Suppose that three possible mechanisms for a reaction with the chemical equation $A + B \rightarrow C$ are as follows:

$$A \rightarrow I_1 \tag{1}$$

$$I_1 + B \rightarrow C \tag{2}$$

and

$$A \rightarrow I_2 \tag{3}$$

$$I_2 + B \rightarrow I_3 \tag{4}$$

$$I_3 \rightarrow C \tag{5}$$

and

$$A \rightarrow I_4 \tag{6}$$

$$I_4 \rightarrow I_5 \tag{7}$$

$$I_5 + B \rightarrow C \tag{8}$$

The I symbols indicate possible intermediate species. Assume that each elementary reaction is "irreversible" and use the steady state assumption to find the rate law for the three cases. Describe the meaning of the result you obtain.

2-10   Consider the general third order reaction with the rate law

$$R = k[A]^{1.0} [B]^{1.0} [C]^{1.0}$$

where the starting concentrations $[A]_0$, $[B]_0$, and $[C]_0$ are all different and the stoichiometric coefficients are all different. Obtain the integrated rate law through application of the partial fraction expansion (cf. Exercise 2-3).

2-11   A certain species of animal has a population of 1 animal per 1 km^2. (a) If the population growth rate is maintained at 10% per 10 years, how long would it take for the population to reach 1 animal per 1 m^2? Assume the population growth follows the mechanism

$$M \rightarrow M + M \tag{1}$$

and use the starting concentration $[M]_0 = 1$ animal/km². (b) This species becomes aggressive when meeting its own kind frequently and destroys itself in a bianimalar reaction

$$M + M \rightarrow C + M \tag{2}$$

with a rate constant $k_2 = 0.5$ m² animal^{-1} year^{-1}. This has the effect of restraining the population explosion caused by reaction (1). Use the rate constant calculated in part (a) and this value of k_2 to determine the steady state value of $[M]$ at which $d[M]/dt = 0$.

2-12 Part of the mechanism for energy production in main sequence stars is

$$^{12}C + H \rightarrow {}^{13}N + \gamma$$

$$^{13}N \rightarrow {}^{13}C + e^+ + \nu$$

$$^{13}C + H \rightarrow {}^{14}N + \gamma$$

$$^{14}N + H \rightarrow {}^{15}O + \gamma$$

$$^{15}O \rightarrow {}^{15}N + e^+ + \nu$$

$$e^- + e^+ \rightarrow 2\gamma$$

$$^{15}N + H \rightarrow {}^{12}C + {}^4He$$

Where e^+ represents a positron, e^- represents an electron, ν represents a neutrino, and γ represents a gamma ray. (a) Show that the stoichiometric equation is

$$4H + 2e^- \rightarrow {}^4He + 2\nu + 7\gamma$$

(b) The reaction consuming ^{14}N is far slower than the rest. Write a rate law for hydrogen consumption treating it as the rate limiting step. (c) Show that the steady state approximation applied to this mechanism predicts that $d[H]/dt = -4k_s\,[{}^{15}O]$ is also a satisfactory rate law.

SUPPLEMENTARY REFERENCES

The separation of variables method used in this chapter to obtain integrated rate laws is not always discussed in introductory calculus courses, although many introductory calculus textbooks do mention it. For example:

J. N. Butler and D. G. Bobrow, *The Calculus of Chemistry* (W. A. Benjamin, Inc., New York, 1965).

W. A. Granville, P. F. Smith, and W. R. Longley, *Elements of Calculus* (Ginn and Co., Boston, 1957), Chapter 21.

Some simple examples of the method are treated in a standard supplementary text:

F. Daniels, *Mathematical Preparation for Physical Chemistry* (McGraw-Hill, New York, 1928), paperback reprint, 1962, Chapter 18.

Additional examples of the procedures used to obtain integrated rate laws are given by

S. W. Benson, *Foundations of Chemical Kinetics* (McGraw-Hill, New York, 1960), Chapters 2 and 3.

A general mathematical procedure that can be used to obtain integrated rate laws is the Laplace transform method. This has been applied to a large variety of rate laws by

N. M. Rodiguin and E. N. Rodiguina, *Consecutive Chemical Reactions* (D. Van Nostrand, Princeton, 1964).

Chapter 3

EXPERIMENTAL METHODS

IN CHEMICAL KINETICS

THERE ARE many kinds of experimental observations that are useful for finding the mechanism of a chemical reaction. In this chapter we survey the principles of the most important ones. Two things in addition to knowledge of these principles can assist in making a wise selection of experiments according to their promise of yielding mechanistic information about a particular reaction. The first is thorough knowledge of the relevant chemical facts; the second is practical experience with the relevant experimental methods.

Like all branches of science, chemical kinetics is a human enterprise undertaken by people working with their tools and their brains. It happens that in kinetics the brains are usually able to keep ahead of the experiments, so that progress is limited by the availability of tools to do more and better experiments. New developments in technology have, therefore, resulted in new understanding of mechanisms and of elementary reactions. For the most part the pertinent new technology has involved more sensitive and more accurate analytical chemistry, provided for example by mass spectroscopy, gas chromatography, and uv–visible–ir spectrophotometry. Modern technology can also provide radically different and powerful kinetics experiments as well, some of which are discussed in Chapter 9. In this chapter we consider methods used in conventional kinetics experiments.

3-1 RATE MEASUREMENTS

From the general form of rate laws and the temperature dependence of rate constants it is clear that the fundamental variables in reaction rate measurements are concentration, time, and temperature. Although it is, in general, relatively simple to measure any of these quantities individually with good or excellent accuracy, the necessity of knowing the values of all three of them together gives rise to measurement problems that can become severe.

Temperatures of chemically reacting systems are usually measured by placing them in an appropriate thermostat and measuring the thermostat temperature with some kind of thermometer, thermocouple, or thermistor. In the temperature range ordinarily used for carrying out chemical reactions, approximately 0–200°C, measuring a thermostat temperature and controlling its value to better than 0.1 deg presents no special problems. Determining the temperature history of a reacting system inside the thermostat, however, is not so simple. At the start of the experiment the system must first come to a thermal steady state with the thermostat, and while the experiment is in progress the heat change caused by reaction must be exchanged with the thermostat. Before accepting a thermostat temperature as the reaction temperature it is necessary to consider these two effects. A simple calculation will often show that these effects may be neglected for the circumstances at hand. This would be the case, for example, for a slow reaction carried out in a dilute, rapidly stirred solution. Knowledge of the reaction heat, reaction time, and thermal conductivity of the system allows one to make an estimate of the degree of temperature equilibration for any kinetics experiment. In some cases it may be necessary to carry out experimental measurements of the temperature history.

If a reaction rate is to be measured at extremely low or high temperatures, thermostat construction becomes more difficult. Fortunately, for rate studies it is not necessary that the temperature be continuously variable. Specific temperatures determined by boiling or melting points of suitable substances can, therefore, be used in reliable thermostats.

In reactions carried out in flowing systems (Section 9-1) the temperature equilibrium problem must be solved together with a mass transfer problem.

Time is also intrinsically easy to measure with excellent accuracy, for example with stopwatches for time intervals of many seconds or with electronic timers for time intervals of a few microseconds. The *reaction time*, however, may be quite difficult to determine. At the start of an experiment the reactants must be mixed, or the premixed reactants must be raised to reaction temperature or otherwise induced to start reacting. None of these are instantaneous events. If the over-all reaction time scale is not very slow compared with the time scale of initiation, the starting error may be significant.

Concentration measurements may be easy or difficult, depending on the substance and the amount to be measured. Whether the pertinent concentration measurement is easy or difficult, the corresponding rate measurement requires knowledge of the concentration of that substance as a function of reaction time. If the concentration can be measured accurately within the system while reaction is occurring, and on a time scale that is fast compared with the time scale for the reaction, then there is no problem here. Spectrophotometric analysis is an example of an analytical method that may have these properties in favorable cases. More often, however, the reaction must be interrupted for analysis. A typical procedure would be to place a system for a measured length of time in a thermostat at high temperature and then "quench" the reaction by cooling the system rapidly. The concentration measurement made later is then assigned to the time at which the reaction was quenched.

The crux of successful rate measurements usually lies in the concentration measurements, for these are the most troublesome of the three principal quantities to measure. If at all possible, a continuous method that does not perturb the system is greatly to be preferred. If this is simply not possible, then care must be taken in relating concentrations determined at leisure to concentrations as functions of reaction time.

Sometimes a physical measurement, such as pressure, refractive index, or temperature is used instead of concentration to determine the extent of reaction. Most of these physical measurements can be made continuously, and they usually involve negligible perturbations of the reacting system. Their main drawback is that it can be awkward to relate the physical property changes to the extent of chemical reaction with sufficient reliability. Where applicable, physical measurements are more convenient than chemical ones.

One over-all view that emerges from considering the measurements of chemical kinetics is that rate measurements, like rate laws, are composite entities, and in designing and evaluating experiments to measure reaction rates the accuracies and limitations of all three components must be considered together. Another previously unexpressed view that you have probably come to suspect for yourself is that rate data are less precise and less accurate that data taken at chemical and physical equilibrium. Whereas statistical analysis of random errors is often merely a matter of good form in other areas of physical chemistry research, it is usually a necessity in chemical kinetics.

3-2 REACTION ORDER DETERMINATION

Since the rate law is one of the most suggestive experimental results for finding a reaction mechanism and since agreement between predicted and experimental rate laws is a necessary condition for every acceptable mechanism, determining the functional form of the rate law is usually considered the first step in a chemical kinetics investigation. Invariably one first tries to use the experimental data to obtain the reaction orders in a simple mass action law. If this proves to be the wrong function, then other functions can be tried. The other functions may reduce to mass action laws for particular limiting conditions, for instance high or low concentration of some reactant, or high or low pressure, or high temperature, or short reaction time.

In this section a number of systematic procedures for determining reaction orders are discussed. It is assumed that a mass action law applies in each case. Failure of these procedures to give consistent results for a reaction is good evidence that the rate law is not of the mass action form. The manner in which these procedures fail in such cases usually indicates alternative functional forms that will fit the data better than a mass action law.

3-2a   Computerized Regression of Concentration/Time Data

Brute force may not have the aesthetic appeal of mathematical insight, but it can be effective where insight fails. The brute in question here is a digital computer. Fast digital computers have become available only in the recent history of chemical kinetics and have therefore not yet played a major role in handling rate data. With the present general availability of computers, however, it is well to ask at the outset of a rate study whether it might be advantageous to turn over all the labor of data reduction to a computer. If the experimental data consist of isothermal concentration versus time measurements for a number of experiments with various starting conditions, as is usually the case, then data analysis by a computer is definitely indicated not only for determining reaction orders, but also for carrying out all of the arithmetic that intervenes between meter readings and writing a report of the results for publication. In this procedure, brainwork is done in writing the computer program at the start and in interpreting the computer output at the finish rather than in doing arithmetic or drawing graphs, which computers can do much faster and much better than people. The bonus of computer data analysis is that full use is made of all the measurements, that the computer makes virtually no mistakes, and that a statistical analysis of the results is automatically provided. In refined experiments computers can even read the meters and control the reaction conditions as well as handle the data.

Let us investigate the problems that arise in a computer analysis of this kind by considering as an example a system with one irreversible chemical reaction of our general form $\sum v_i A_i = 0$. Let the set of starting concentrations be $\{[A_i]_0\}$, let the desired set of reaction orders be $\{n_i\}$, and let x be the change in concentration of the substance A_1 whose concentration is measured as a function of time in the experiments. Then

$$[A_i] = [A_i]_0 - \frac{v_i}{v_1} x$$

We assume a rate law of the mass action form

$$v_1 R \equiv \frac{-dx}{dt} = v_1 k(T) \prod_{\text{reactants}} [A_i]^{n_i} = v_1 k(T) \prod_{\text{reactants}} \left([A_i]_0 - \frac{v_i}{v_1} x \right)^{n_i}$$

This may be rearranged to give

$$\frac{dx}{\prod\limits_{\text{reactants}} \left([A_i]_0 - \frac{v_i}{v_1} x \right)^{n_i}} = -v_1 k(T)\, dt$$

$$\sum_{\text{reactants}} \frac{c_i\, dx}{\left([A_i]_0 - \frac{v_i}{v_1} x \right)^{n_i}} = -v_1 k(T)\, dt$$

$$\sum_{\text{reactants}} \frac{c_i/v_i}{1 - n_i} \left([A_i]_0 - \frac{v_i}{v_1} x \right)^{1 - n_i} - [A_i]_0^{1-n_i} = k(T)\, t$$

The set $\{c_i\}$ is the set of coefficients in the partial fraction expansion; they are found as functions of the sets $\{[A_i]_0\}$ and $\{v_i\}$ by the partial fraction technique. (For a summary of the mathematics of simple partial fractions see Exercise 2-3.)

We seek the $\{n_i\}$ that will give the best rate law for a large set of data points x_j, t_j, $\{[A_i]_0\}_j$, that is, for the A_1 concentrations at various reaction times for experiments with different starting concentrations. A conventional least squares analysis on the above integrated rate equation looks hopeless: but brute force does not. The computer need only be asked to compute as many values of $k(T)$ as there are data points, which could be indexed as $k_j(T)$ for a succession of different assumed sets $\{n_i\}$. One might choose to compute $\{k_j(T)\}$ sets for all combinations of n_i with n_i in the range $-1.0 < n_i < 2.0$, taking care to avoid 1.00000, in increments of 0.2. If orders with respect to three substances are sought, the number of $\{k_j(T)\}$ sets to be calculated is then a trifling 4096. Each set can be tested for consistency by forming the residual sum $\sum (k_j(T) - \langle k_j \rangle)^2$, where $\langle k_j \rangle$ is the average $k(T)$ for a given $\{n_i\}$; the $\{n_i\}$ sets showing the smallest residual sums will indicate the approximate best reaction orders. Repeating the calculation for smaller increments near the approximate orders will provide the reaction orders to all the accuracy permitted by the measurements themselves. Once a rate law has been found in this way, it can easily be subjected to statistical tests by finding the deviations between the data points and their corresponding calculated $x(t)$ values.

This procedure was presented in detail to show that a rate law can be derived by straightforward computing, not to give a formula to be used for a broad variety of practical applications.   Every rate study will have its own computing peculiarities: one might wish to look for different rate laws for different conditions, set some orders at particular values while searching for others, assign different weights to different measurements, try rate laws not of the mass action form, and so on.   Using a computer to do arithmetic frees the brain for these more worthwhile efforts.

3-2b   Initial Rates

The course of most chemical reactions is simplest at the very beginning, before enough products accumulate to react among themselves and before effects of decreasing reactant concentrations appear.   If analytical methods sensitive enough to determine rates at small fractions of reaction are available, then determining reaction orders is greatly simplified.   All reactant concentrations are simply set equal to their starting values, and the initial rate law becomes

$$R_{\text{initial}} = k(T) \prod [A_i]_0^{n_i}$$

One holds the $[A_i]_0$ constant for all but one substance, measures the dependence of R_{initial} on that $[A_i]_0$ to find the corresponding n_i, then repeats for all A_i.

Studying initial rates can be a simple and reliable procedure.   Where feasible it is highly recommended.

3-2c   Isolation

This is another way of finding reaction orders individually.   One arranges that all reactants save one, for instance, A_k, are present in large concentrations. For them $[A_i] = [A_i]_0$ throughout the reaction, and the variation with time of the remaining $[A_k]$ should give the reaction order with respect to this A_k.

An unfortunate corollary of the isolation idea is that if one reactant is always present in large excess it is not possible to determine a reaction order with respect to it.   This is always the case for the solvent in solution reactions, since its concentration is usually about 30 to 60 molar.

3-2d   Graphs

Often the expense of reagents or lack of time preclude doing many rate experiments on a reaction.   If the stoichiometry of the reaction is simple, then graphical methods may be used to determine orders with fair accuracy from a few or even only one experiment.   The graphs of $[A]$, $\log[A]$, $[A]^{-1}$ and

$[A]^{-2}$ versus time introduced in Chapter 2 are the most frequently used ones. In any particular case, we have some function obtained by integrating an assumed rate law, and the data points are substituted into it for plotting versus reaction time. If a line can be drawn through the data to within the known experimental errors, then it may be concluded that the assumed rate law has the correct reaction order or orders in it.

A number of specialized graphical methods are described in advanced treatises on chemical kinetics.

3-2e   Rate Constant Consistency

Another procedure that uses integrated assumed rate laws is sometimes helpful for approximate order determinations. A collection of $[A_i]_t$, t data points is substituted into one integrated rate law after another to calculate sets of rate constants. Then the several sets of rate constants are compared with one another. The most nearly correct rate law should be the one for which the set of calculated rate constants shows only random scatter about the average value, with the other rate laws having rate constant sets that change systematically with starting concentrations and reaction times.

If all of the trial integrated rate laws show systematic rate constant changes, then other functional forms should be tried. The manner in which the trials fail will indicate appropriate alterations.

3-2f   Fractional Lives

There is no mathematical reason for always considering time to be an independent rather than a dependent variable. Therefore we may invert integrated rate laws and determine reaction orders by measuring the dependence upon starting concentrations of the reaction time required for a given fractional decrease in starting concentration of reactions that are zero, first, second, or third order in one substance. It may be shown (Exercise 3-3) that these reaction times are proportional to the first, zeroth, minus one, and minus two powers, respectively, of the starting concentration of that substance. If it is convenient to measure fractional lives in a reaction as a function of starting concentration, the appropriate log–log graphs will allow the reaction order to be found.

3-3   RATE CONSTANT REPORTING

Reaction orders computed from experimental data by the above methods usually are nearly, but not quite, within experimental error of the integral values 0, 1, or 2. If the rate law is only needed for computing reaction rates, for example, for preparative or industrial purposes, then there is no objection to reporting the rate law just as it is obtained. As an example, the expression $R = 4.2 \times 10^3[A]^{0.46}[B]^{1.10}$ moles liter^{-1} sec^{-1} is useful for such purposes. On the other hand, it is sometimes the case—particularly if one believes that the rates measured are, or are related to, elementary reaction rates—that one wishes to report a rate law with integral or half-integral orders. In the example just given this would be $R = k(T)[A]^{0.5}[B]^{1.0}$. If this is done, then the value of the rate constant must be recalculated. Normally it will be significantly, although not drastically, different from the rate constant calculated when the orders are derived directly from experimental data.

First order rate constants have the dimensions of $[\text{time}]^{-1}$. Except for radioactive decay rate constants, which are usually reported as half-lives anyway, *seconds* are generally chosen as the time units for expressing the value of the rate constant. Rate constants for all other reaction orders have the dimensions of inverse time multiplied by concentration raised to the power $(1 - n)$, where n is the overall reaction order. The choice of concentration units is a matter of personal preference. The most popular are *moles/liter, moles cm^{-3}*, and *molecules cm^{-3}*. Since it is customary in physical science to omit naming the unit of particles (i.e., atoms, pions, molecules) occurring in composite quantities, molecules cm^{-3} units often are expressed simply as cm^{-3}; a second order rate constant for molecules cm^{-3} concentration units will have *cm^3 sec^{-1}* units in this book. It is not unusual to find rate constants reported in special-purpose units such as torr^{-1} min^{-1}, kg^{-1} day^{-1}, or μsec^{-1} atm^{-1} in specialized areas of chemical kinetics.

Rate laws with nonintegral over-all order have rate constants with concentration units raised to nonintegral powers. In the example given above, the rate constant 4.2×10^3 has units liters$^{+0.56}$ moles$^{-0.56}$ sec^{-1}.

In this book we express concentrations as molecules per cubic centimeter or moles per liter and time in seconds since these units are familiar to chemistry students. Some of the problems use different units to provide practice in conversion from one set of rate units to another.

3-4 *ACTIVATION ENERGY DETERMINATION*

The chemical changes that occur in a system of given starting composition usually vary markedly with temperature. Therefore, one would expect that reaction rates may vary with temperature in different, complex ways depending on what reaction mechanisms predominate in various temperature ranges. If a single reaction mechanism does characterize a system of interest over a substantial range of temperature, however, then the temperature dependence of the reaction rate can be described by a simple function. This will almost certainly be the case if the conversion rate is determined by a single elementary reaction rate. In most cases the temperature dependence of rate constants can be taken from the Arrhenius form

$$k(T) = A \exp(-E_A/RT)$$

to within the accuracy of the experimental data. For the few cases where the Arrhenius function does not fit the experimental data, a function that does is

$$k(T) = BT^m \exp(-E_B/RT)$$

The parameters of these two functions are related through the equations (cf. Exercise 3-17).

$$A = B(eT)^m$$

$$E_A = E_B + mRT$$

where e is the base of natural logarithms.

Sometimes E_B is set equal to zero, giving

$$k(T) = BT^m$$

In Chapters 4 and 5 we find that each of these three $k(T)$ functions can be given theoretical meaning.

To avoid ambiguity we use the term "activation energy" *only* to denote E_A, which is then by definition $(-R)$ times the slope of the graph of $\ln k(T)$ versus $(1/T)$

$$E_A \equiv -R \frac{d \ln k(T)}{d(1/T)}$$

It proves convenient to use the term "preexponential factor" interchangeably for the constants A and B in the above equations; as long as the value of m is known from the context there is no ambiguity here either. These distinctions are important when discussing theoretical $k(T)$ functions for elementary reactions since B, m, and E_B are then expressed in terms of molecular parameters.

In complete contrast to the situation with reaction order determination, determining the parameters in a $k(T)$ expression is a straightforward matter. The first step is to collect the data on a graph of $\log k(T)$ versus $(1/T)$. The second is to draw a straight line through the data. We can tell at a glance whether the precision of the data warrants fitting a $k(T)$ expression with pre-exponential temperature dependence. Almost always it does not. Then we merely calculate $\log A$ and $E_A/2.3R$ as parameters of the straight line equation

$$\log k(T) = \log A - E_A/2.3RT$$

using the linear least squares equations.

Often a chemist believes that his rate data give an elementary reaction rate and wishes to have an experimental $k(T)$ function with a particular theoretical value of m for comparison with a theoretical $k(T)$ functon. The procedure then is to use the linear least squares equations to find the best values for $\log B$ and $E_B/2.3R$ in the straight line equation

$$\log \frac{k(T)}{T^m} = \log k(T) - m \log T = \log B - \frac{E_B}{2.3RT}$$

If an experimental value for m is desired, a nonlinear least squares computer program can be used. If this is not available, then one should first divide the data into " high T " and " low T " groups and estimate two values of E_A from them graphically. By forming the second derivative of $k(T)$ we find

$$\frac{d^2 \ln k(T)}{d(1/T)^2} = \frac{d^2}{d(1/T)^2} \left[\ln B + m \ln T - \frac{E_B}{RT} \right] = mT^2$$

In the Arrhenius form the first derivative is

$$\frac{d \ln k(T)}{d(1/T)} = -\frac{E_A}{R}$$

The estimated second derivative would be

$$\frac{-E_A(\text{low } T)/R + E_A(\text{high } T)/R}{1/\text{low } T - 1/\text{high } T}$$

$$= \frac{(\text{high } T)(\text{low } T)}{R} \frac{E_A(\text{high } T) - E_A(\text{low } T)}{\text{high } T - \text{low } T}$$

Equating mT^2 with the estimated value of the second derivative gives

$$m = \frac{(\text{high } T)(\text{low } T)}{RT^2} \frac{(E_A(\text{high } T) - E_A(\text{low } T))}{(\text{high } T - \text{low } T)}$$

$$\cong \frac{E_A(\text{high } T) - E_A(\text{low } T)}{R(\text{high } T - \text{low } T)}$$

as a preliminary value.   Starting with this and nearby values for m we can obtain more accurate values for m by repeated trials with the same linear least squares equation used for known m.

Most deviations from the Arrhenius $k(T)$ function prove to be due to changes in the mechanism with temperature.   The determination of meaningful experimental m values for elementary reactions is mostly restricted to solution reactions.   For the few gas reactions where m has been found experimentally, E_B was assumed known on theoretical grounds and m could be determined from the form

$$\log k(T) + E_B/2.3RT = \log B + m \log T$$

The units used for reporting E_A and E_B are almost invariably kilocalories.

Special $k(T)$ functions are occasionally used for some reactions involving H-atom or proton transfers where quantum mechanical tunneling is thought to contribute to the reaction rate.

3-5   CHEMICAL KINETICS AND THERMOCHEMISTRY

We noted in Chapter 1 that the thermodynamic functions $A(\xi)$ and $G(\xi)$ provide information about the direction of possible chemical reactions.   It is convenient at this point to consider some further interconnections between reaction rates and thermodynamics.   Briefly, chemical kinetics is indebted to thermodynamics for some useful guidelines concerning the magnitudes of activation energies and preexponential factors, and repays this debt by providing values for important thermochemical quantities that are difficult or impossible to obtain by other means.

Let us denote the forward and reverse rate constants of an elementary reaction by $k_f(T)$ and $k_r(T)$, respectively.   We showed in Chapter 2 that $k_f(T)/k_r(T) = K_{eq}$.   Let us specialize to the case of an ideal gas reaction with $\sum v_i = 0$, so that $K_{eq} = K_p$ and $\Delta H = \Delta E + \Delta(pV) = \Delta E$.   Then the thermodynamic relation $K_p = \exp(-\Delta G^0/RT)$ gives

$$k_f(T)/k_r(T) = K_{eq} = K_p = \exp(-\Delta G^0/RT)$$

$$= \exp(-\Delta H^0/RT)\exp(\Delta S^0/R)$$

$$= \exp(\Delta S^0/R)\exp(-\Delta E^0/RT)$$

In Arrhenius form the rate constant ratio is

$$\frac{k_f(T)}{k_r(T)} = \frac{A_f \exp(-E_{Af}/RT)}{A_r \exp(-E_{Ar}/RT)}$$

The thermodynamic quantities ΔS^0 and ΔE^0 are functions of temperature. They vary with temperature far more slowly than $\exp(-\text{constant}/T)$, however, and to the accuracy with which virtually all kinetics experiments are done they may be regarded as constants. Let us consider the entire temperature variation to be the explicit $1/T$ in the exponential factors and take ΔS^0, ΔE^0, A_f, A_r, E_{Af}, and E_{Ar} to be constants. Comparing the two equations for $k_f(T)/k_r(T)$—or their logarithmic forms—then leads to the results

$$\boxed{\begin{aligned} A_f/A_r &= \exp(\Delta S^0/R) \\ E_{Af} - E_{Ar} &= \Delta E^0 \end{aligned}}$$

The relationship between the activation energies for the forward and reverse reactions is illustrated in Figure 3-1.

EXERCISE 3-1 Rederive these results for $k(T) = BT^m \exp(-E_B/RT)$ and $\Delta \nu = 0$, but still for an ideal gas reaction.

These two equations have a number of implications. The first is that measuring the parameters A and E_A for one direction of an elementary reaction suffices for finding the same parameters for the reaction in the reverse direction as well if thermochemical quantities ΔS^0 and ΔE^0 are known, as they usually are. This has the effect of doubling the rate of data accumulation on elementary reactions of known thermochemistry.

On the other hand, if ΔS^0 and ΔE^0 are not known, measurement of the A and E_A parameters for both forward and reverse directions will provide experimental values for them. For some reactions ΔS^0 can be calculated from spectral data; then measuring k_f and k_r at a single temperature gives ΔE^0. The most important applications of this idea are in deducing heats of formation and bond dissociation energies of unstable species. For example, from the A and E_A parameters for the elementary reaction

$$Br + CH_4 = HBr + CH_3$$

it was possible to deduce $\Delta H_f^0(CH_3) = 33$ kcal/mole and $D(CH_3—H) = 102$ kcal/mole, with the values of $H_f^0(Br)$, $H_f^0(CH_4)$, and $H_f^0(HBr)$ being known from spectroscopic and calorimetric experiments. The principal limitation of this method of obtaining thermochemical data, aside from the reduced accuracy compared to calorimetric data, is that the elementary reaction concerned is necessarily part of a mechanism, and the derived thermochemical results are only as reliable as the mechanism assumed in deriving the A and E_A data for its component elementary reaction. Balancing this limitation is the

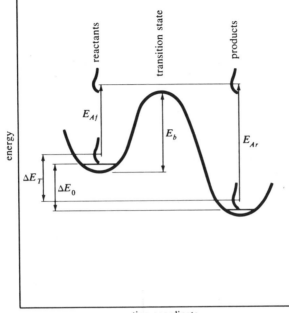

reaction coordinate

FIGURE 3-1 *Schematic diagram illustrating relationships between activation energies and thermodynamic quantities for a chemical reaction in which the mechanism and the stoichiometric equation are identical. E_{Af} and E_{Ar} denote the activation energies of the forward and reverse reactions; ΔE_T is the heat of reaction (but see comment below) at temperature T; ΔE_0 is the heat of reaction at $0°K$; E_b is the barrier height (cf. Section 4-3). From the graph it is clear that $\Delta E_T = E_{Af} - E_{Ar}$. The reaction coordinate denotes some measure of the change in molecular configuration that accompanies transformation of reactant molecules to product molecules (cf. Section 4-7). Since ΔE_T usually varies with temperature, because of the difference in heat capacities of the reactants and the products, E_{Af} and E_{Ar} would be expected to be temperature dependent also. However, observed changes of E_A with temperature are usually caused by changes in mechanism and stoichiometry instead. The curve represents here the potential energy variation in a single encounter as the reactants form first a transition state and then products (cf. Section 4-3). For a mechanism with several elementary reactions the curved line would have a succession of maxima corresponding to the various transition states. The quantity ΔE_T used here is exactly equal to the change in internal energy only for a chemical reaction involving only ideal gases. In thermodynamics, the term "heat of reaction" usually refers to the enthalpy change ΔH_T when a chemical reaction proceeds from $\xi = 0$ to $\xi = 1$ at temperature T; if the reaction is carried out at constant pressure and no work other than pV work is done, the heat of reaction is also the actual heat transferred from the surroundings to the system during the reaction. The heat of reaction ΔE_T obtained from $E_{Af} - E_{Ar}$ is nearly, but not identically, equal to the enthalpy change ΔH_T for most reactions.*

fact that the two other ways of obtaining bond dissociation energies have limitations themselves: spectroscopic methods are useful only for small molecules, and electron bombardment methods also involve mechanisms.

Very few elementary reactions are known to have negative values of E_A. (These cases are discussed in Chapter 5.) This implies that for an endothermic elementary reaction E_A is usually at least as large as the endothermicity of the reaction ΔE_T. Therefore, for estimating the values of rate constants of endothermic reactions we have a lower limit for E_A available if ΔE^0 for the reaction is known.

It has long been hoped by chemists that reaction rates might be predictable from thermodynamic data and the structures of the reacting molecules. Indeed, the ultimate aim of chemical kinetics could almost be phrased in these terms alone as the explanation and prediction of molecular reactivity from the properties of nonreacting molecules. A part of this aim has been to find relationships between activation energies and heats of reaction. For an elementary reaction proceeding in the exothermic direction, it seems plausible that higher activation energies ought to be required if strong bonds are broken than if weak bonds are broken; alternatively, lower activation energies ought to be required if strong bonds rather than weak bonds are formed. A simple function expressing this idea for a family of closely related reactions would be a linear equation

$$E_A = \alpha \Delta E^0 + \text{const}$$

where α is an empirical proportionality constant. Unfortunately, this and other attempts are not yet generally successful for gas reactions, although there is considerable success for solution reactions (Section 6-4). Part of the difficulty is that nature tends to ignore the plausible assumption. For example, comparing the almost thermoneutral elementary gas reactions

$$D + H_2 = HD + H$$

$$^{16}O + {}^{18}O_2 = {}^{16}O—{}^{18}O + {}^{18}O$$

we find that the respective activation energies are 9 and less than 3 kcal, whereas the dissociation energies of the bonds to be broken in each reaction are 103 and 118 kcal, respectively. Many such examples could be given. It is quite clear that other factors besides bond strengths contribute to activation energies. In spite of difficulties in understanding what these factors are, it is to be expected that satisfactory empirical relationships connecting A and E_A parameters to thermochemical and other molecular data will eventually be found. Further consideration is given to this point in Chapter 4.

Our discussion of the interrelation between thermodynamics and chemical kinetics has focused on gas reactions. For solution reactions the same basic considerations apply and the same formulas are derived. Deriving the final equations, however, is somewhat more complicated for solution reactions.

3-6 EVERYTHING ELSE THAT PERTAINS

The conscientious investigator of reaction mechanisms will try every experiment he can think of that promises to give clues to or confirmation of the elementary reactions taking place in the system in which he is interested. We consider in this section the principles underlying the most important experiments available for this purpose. The technology, which is largely nothing more than careful analytical chemistry, we omit.

3-6a Product Analysis

Obviously we must know what products are formed in a chemical reaction before we know what any elementary reaction might be. What is not so obvious is the importance of a *quantitative* analysis for *all* of the reaction products. During most of the history of chemical kinetics this was too great a demand upon a chemist due to the limited analytical methods that were available. This is, however, no longer the case. Since most chemical reactions give more complex arrays of products as the reaction time is increased, it is preferable to carry out a complete analysis for the shortest possible reaction time, or better yet to find the product distribution for a range of short reaction times. In this way one can differentiate between primary products formed directly from reactants and secondary products formed from further reactions of primary products. The identity of some of the minor products may give valuable clues to the identity of intermediate species leading to the major products. The mechanism eventually proposed must account for the observed product distribution as well as for the observed rate law.

3-6b Isotopes and Tracers

Additional insight into reaction pathways is possible if one can discover which parts of the reactant molecules end up as which parts of the product molecules. Normally the chemical reactivity of isotopic atoms such as ^{13}C, ^{15}N, or ^{18}O is not significantly different from that of the isotopes which predominate in natural abundance, that is ^{12}C, ^{14}N, or ^{16}O. A large variety of compounds with certain atomic positions isotopically labeled with supernatural abundance of one isotope has been synthesized. These compounds are available commercially for use primarily in mechanism studies.

There is a threefold choice of analytical method for isotope experiments, namely, mass spectroscopy, radioactive tracing, and magnetic resonance. The method of choice in a particular problem is largely determined by the particular reaction and the particular compounds involved. Mass spectroscopy is generally the simplest experimentally; radiochemical methods are the most sensitive and therefore require the least reaction times, but one must take the trouble of setting up a radiochemical laboratory; magnetic resonance gives information about the location of the isotopic atoms in the product molecules, but requires large amounts of sample. The method of high resolution mass spectroscopy promises to become the most powerful method of all, for it combines high sensitivity with ability to determine the isotopic percentages at all atomic positions in reactant and product molecules. Each of these methods of determining the fates of labeled molecules is considerably enhanced by combining them with vapor phase chromatographic separation of the compounds in the product mixtures before the molecules are analyzed according to the nuclear properties of their atoms.

The mass ratio of hydrogen and deuterium is large ($m_D/m_H = 2$) compared with, for example, the mass ratio $13/12 = 1.08$ for the stable carbon isotopes. Elementary reactions that involve breaking bonds of hydrogen atoms have different rates than the same elementary reactions breaking corresponding bonds to deuterium atoms. This is called the *primary kinetic isotope effect*. A semiquantitative explanation is as follows. Suppose that a hydrogen atom is bonded to a heavy radical R. The reduced mass governing the vibration frequency of the R—H stretch is $\mu_{RH} = m_R m_H/(m_R + m_H) \cong m_H$, since $m_R \gg m_H$. The R—H stretching frequency is $\nu_0 = (C/\mu_{RH})^{\frac{1}{2}}(2\pi)^{-1}$, where C is the force constant of the R—H bond, and the zero-point energy is $\varepsilon_{0H} = \frac{1}{2}h\nu_0 = \frac{1}{2}\hbar(C/\mu_{RH})^{\frac{1}{2}} \cong \frac{1}{2}\hbar(C/m_H)^{\frac{1}{2}}$, where $\hbar = h/2\pi$. The zero-point energy becomes $\frac{1}{2}\hbar(C/m_D)^{\frac{1}{2}}$ if the hydrogen atom is replaced by a deuterium atom. Let us now suppose that the R—H or R—D bond is stretched in some elementary reaction until the H or D atom is transferred to some other molecule. The "activation" energy required to stretch R—H by a given (large) amount will be less than the energy required to stretch R—D by the same amount due to the different zero-point energies from which the stretching starts, as shown in Figure 3-2. The relative rate constants are given by

$$\frac{k(T)_{R-H}}{k(T)_{R-D}} = \frac{A \exp(-(E_{\text{stretch}} - \varepsilon_{0H})/RT)}{A \exp(-(E_{\text{stretch}} - \varepsilon_{0D})/RT)}$$

$$= \frac{\exp(\varepsilon_{0H}/RT)}{\exp(\varepsilon_{0D}/RT)}$$

$$= \frac{\exp(\frac{1}{2}h(C/m_H)^{\frac{1}{2}}/RT)}{\exp(\frac{1}{2}h(C/m_D)^{\frac{1}{2}}/RT)}$$

The rate constant ratio is therefore given by a ratio of exponentials of the zero-point energies divided by RT. For typical saturated organic molecules the C–H stretching modes appear at 3–3.5 μ in the ir spectrum. If we take 3.3 μ as typical, then $\varepsilon_{0H} = \frac{1}{2}h\nu_0 = \frac{1}{2}hc/\lambda \simeq (0.5)(6.6 \times 10^{-27})\,(3.0 \times 10^{10})/$ $3.3 \times 10^{-4} = 3.0 \times 10^{-13}$ ergs/molecule = 4.4 kcal/mole. For ε_{0D} the same estimate gives 3.1 kcal, since the force constant for R—D is the same as for R—H. At a reaction temperature of 373°K the rate constant ratio is

$$\frac{k(T)_{R-H}}{k(T)_{R-D}} = \frac{\exp[4{,}400/(1.99 \times 373)]}{\exp[3{,}100/(1.99 \times 373)]} \cong 5.7$$

The primary kinetic isotope effect thus gives rise to large changes in reaction rates when deuterium atoms are displaced rather than hydrogen atoms. This is obviously a serious complication for labeling experiments. At the same time it provides an additional opportunity for the study of mechanisms if the rate-determining elementary reactions involved are H-atom or proton transfers.

EXERCISE 3-2 Calculate the relative rates of breaking CH_3—CH_3 and CD_3—CD_3 bonds using assumptions similar to those presented above. The C—C stretching frequency in ethane is 2.98×10^{13} sec^{-1}.

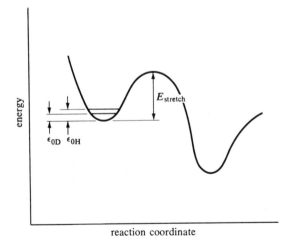

reaction coordinate

FIGURE 3-2 *Schematic energy diagram illustrating the primary kinetic isotope effect. The energy $E_{stretch}$ required to stretch the R—H or R—D bond takes the place of the barrier height on the general energy diagram shown in Figure 3-1. On this simplified diagram the corrections for thermal energy of the reactants and the transition state are not shown; in particular the zero point energy of the transition state is ignored. Allowing for this would decrease the magnitude of the primary kinetic isotope effect.*

3-6c   Initiators and Scavengers

We have noted before that most elementary reactions in chemistry involve intermediate species that are not isolable compounds. Detecting such species directly is difficult, although not impossible (cf. Chapter 9). Indirect methods are more readily available. We can divide them into those in which intermediates are added by means of *initiators* and those in which intermediates are removed by means of *scavengers*. Whereas the product analysis and labeling experiments tell what goes where, perturbing a system by adding initiators and scavengers can tell which routes are open for going from here to there.

The intermediates generated by initiators are compounds with unsatisfied or "free" valences called *free radicals* or *radicals*. By generating a known type of radical in a system at a known rate, one can judge the possible participation of the same or other types of radical in the system under study when the initiator is absent. Scavengers can be used in several ways. The simplest is to observe whether the conversion rate or the product distribution is affected upon addition of a substance known to be highly effective in removing certain intermediates. The most elegant is to "trap" the intermediates by forming an isolable, stable compound from them. Since quantitative studies of initiation and scavenging are most prominent in research on chain reactions (Chapter 8) and photochemistry (Chapter 10), we defer the discussion of their use until later.

3-6d   Stereochemistry

On the basis of a product distribution and a rate law alone we can often develop a fairly detailed set of elementary reactions. The postulated elementary steps of reactions of inorganic complexes or larger organic molecules may be tested by a study of the stereochemistry, which may also provide additional details of the transition between reactant and product molecules. In some cases the stereochemical experiment is straightforward; for instance, we can investigate whether optical activity is retained in a given reaction. A classic example involving geometrical isomers is the case of bromination of cyclic alkenes. Formation of the *trans* adduct rather than the *cis* adduct shows that the Br_2 molecule does not add to the olefinic bond in a single elementary reaction. In most stereochemical investigations of mechanisms, working out the stereochemistry of the over-all transformation can be a long, painstaking task demanding far more effort than finding the rate law and product distribution. Since advanced courses in organic and inorganic chemistry devote large amounts of time to this topic, we shall discuss it only lightly in this book.

3-6e   Comparison Reactions

Although each chemical reaction is an individual, it is very likely to be closely related to some of the members of a large set of reactions. We would expect, for example, that most reactions of diethylamine are not drastically different from the corresponding reactions of dimethylamine or di-*n*-propylamine. Research in chemical kinetics is influenced in several ways by similarities in reactivity among related compounds.

Before starting a laboratory investigation of a reaction, the published results of previous investigations on related reactions should be studied. A reasonable guess can then usually be made about what general observations can be expected and what general type of mechanism will rationalize them. The degree to which one can be led astray in starting out with wrong preconceived ideas is generally inconsequential compared with the saving of effort that can be achieved by a study of existing information. A second benefit of a literature survey is that the large reservoir of existing ideas about elementary reactions contains many valuable stimuli to the imagination that can help in sorting out one's own ideas about the mechanism of a particular reaction.

The fact that related compounds often react in similar ways implies that there are similarities in their reaction mechanisms. Such similar mechanistic interpretations support one another in much the same way as scientific theories about all aspects of nature form interwoven, self-supporting rather than externally supported, sectors of knowledge. Therefore, it is risky to propose a mechanism that contains elementary reactions vastly different from those accepted for related reactions. It is important to realize, however, that apparent failures of apparently comparable reactions to be explicable with comparable mechanisms may be a valuable clue to improved understanding of them all.

The practical implication of all this is that it is advisable to become familiar with the pertinent chemical literature before investigating mechanisms on one's own.

3-7   FROM OBSERVATIONS TO MECHANISMS

A chemist engaged in a kinetics research project will have started his work with preconceived ideas about the mechanism of the reaction under study, for he will know the main products of the reaction and the general chemistry involved before doing any kinetics experiments at all. As he proceeds to find the reaction orders, the complete product distribution, the effects of changes in temperature, pressure, solvent, or whatever he feels will be relevant, his ideas about the mechanism mature as they are shuffled, extended, discarded, renovated, and finally assembled into a coherent mechanism. Each mechanism study evolves in its own more or less chaotic way, with the chemist involved doing more or less systematic experiments to make mechanistic order out of the apparent chaos. By the time his experiments are finished, his ideas about the reaction mechanism will have resolved themselves into a fully coherent form, and the coherence of mechanism and experimental results will give him the satisfying feeling that he understands the mechanism of the reaction he set out to study. Then it is time to contribute this understanding to the chemical public. In the publication, the ideas and results that evolved chaotically will be presented in an orderly, logical way. To the uninitiated, this presentation conveys the false impression that the research itself was actually done in this orderly, logical way. To the initiated, however, this presentation is a final ordering and logical rationalization of the experimental findings; it is intended to be an efficient means of communicating the results of the kinetics investigation.

We cannot begin to describe here the vicissitudes of mechanism investigations, for they are as individualistic as are chemists themselves. We can, however, give some skeleton examples of final presentations.

The following examples are selected from reactions whose mechanisms happen to be simple. The experimental results quoted mostly represent the combined efforts of several investigators of each reaction. More detailed discussions, as well as references to the original literature, will be found in the supplementary references listed at the ends of later chapters of this book.

Iodination of acetone. Acetone reacts with iodine in acidic aqueous solution according to the stoichiometric equation

$$CH_3\text{---}CO\text{---}CH_3 + I_2 \rightarrow CH_3\text{---}CO\text{---}CH_2I + HI$$

The product analysis shows that iodoacetone and hydroiodic acid are the only products for short reaction times. The conversion rate for short reaction times was found to be first order in acetone, first order in hydrogen ion, and zero order in everything else. It appears that the rate controlling step is the formation of an intermediate in an elementary reaction between H^+ and CH_3COCH_3, which then reacts rapidly with iodine to form the observed products. The postulated intermediate is the enol form of acetone, giving the mechanism

$$CH_3\text{---}CO\text{---}CH_3 + H^+ \xrightarrow{\text{slow}} CH_3\text{---}COH\text{=}CH_2 + H^+$$

$$CH_3\text{---}COH\text{=}CH_2 + I_2 \xrightarrow{\text{fast}} CH_3\text{---}CO\text{---}CH_2I + HI$$

According to this mechanism the elementary reaction whose rate is equal to the conversion rate is the enolization step. Strong evidence in favor of this idea is provided by the finding that the bromination and iodination of acetone have identical rates. An elaboration of the mechanism would show that the H^+-catalyzed enolization step consists of proton transfer from H_{aq}^+ to the carbonyl oxygen followed by a second proton transfer from a methyl group to a nearby water molecule.

$CO_2^{\ddagger}$ chemiluminescence. Blue light is produced when oxygen atoms react with carbon monoxide according to the stoichiometric equation $O + CO \rightarrow CO_2$. The intensity of the light (quanta cm^{-3} sec^{-1}) was found to follow the rate law

$$R = k_M[0]^{1.0}[CO]^{1.0}.$$

The M subscript indicates that the emission intensity was found to depend upon the nature of any foreign gases present (the symbol M is used to denote molecules that act as "energy-exchanging collision partners" in gas reactions; cf. Chapter 5), whereas the absence of a [M] factor in the rate law indicates that the foreign gas pressure did not influence the reaction rate. A mechanism that rationalizes this rate law is

$$CO + O + M \xrightarrow{k_1} CO_2^{\ddagger} + M$$

$$CO_2^{\ddagger} + M \xrightarrow{k_2} CO_2^* + M$$

$$CO_2^{\ddagger} \xrightarrow{k_3} CO_2^* + h\nu$$

$$CO_2^* + M \xrightarrow{k_4} CO_2 + M$$

The termolecular step is rate limiting. Application of the steady state assumption shows that the observed rate law implies $k_2[M] \gg k_3$, that is, most of the $CO_2^{\ddagger}$ molecules do not emit light on their way to CO_2^*. The species CO_2^* represents a CO_2 molecule that has a high degree of vibrational excitation; it is required in the mechanism in order to account for the fact that CO_2 does not absorb blue light.

Synthesis of urea. One of the milestones of organic chemistry was the discovery that the inorganic substance ammonium cyanate reacts when heated to form the organic substance urea

$$NH_4^+ + CNO^- \rightarrow (NH_2)_2CO$$

By measuring the effects of KCNO and $(NH_4)_2SO_4$ additions upon the conversion rate, the rate law was found to be

$$R = k[NH_4^+]^{1.0}[CNO^-]^{1.0}$$

and on this basis the mechanism was judged to be identical to the stoichio-metric equation.　It was noted later, however, that the acid–base equilibria

$$HCNO \overset{K_a}{\rightleftarrows} CNO^- + H^+$$

$$NH_3 + H_2O \overset{K_b}{\rightleftarrows} NH_4^+ + OH^-$$

are established in elementary reactions that are much faster than urea synthesis.　If these two reactions are in equilibrium, their sum

$$NH_3 + HCNO(+ H_2O) \overset{K}{\rightleftarrows} (H^+ + OH^-) + NH_4^+ + CNO^-$$

is also in equilibrium.　Therefore the mechanism might instead consist of the rate limiting step

$$NH_3 + HCNO \overset{k'}{\rightarrow} (NH_2)_2CO$$

plus whatever elementary reactions maintain the acid–base equilibria.　Since

$$[NH_4^+][CNO^-] = K[NH_3][HCNO]$$

the conversion rate for this alternate mechanism is

$$R = k'[NH_3][HCNO]$$

$$= \frac{k'}{K}[NH_4^+][CNO^-]$$

Both mechanisms thus imply the observed rate law.　There is no way of distinguishing between them by rate measurements or by theories of reaction rates.　Since the lone pair of electrons on NH_3 is available for forming an N—C bond to HCNO, the second mechanism appears more likely.

Phosgene synthesis and decomposition.　The reaction

$$CO + Cl_2 \rightarrow COCl_2$$

is one of the infrequent reactions for which rate measurements can be made in both the forward and the reverse directions.　The conversion rate in the forward direction was found to be described at short reaction times by

$$R_f = k_f[Cl_2]^{1 \cdot 5}[CO]^{1 \cdot 0}$$

The reverse reaction, phosgene decomposition, has the rate law

$$R_d = k_d[Cl_2]^{0.5}[COCl_2]^{1.0}$$

The half-integral orders with respect to chlorine provide a valuable hint about the mechanism. If the equilibrium

$$Cl_2 \overset{K}{\rightleftarrows} 2Cl$$

is maintained by elementary reactions that are rapid compared to those which involve $COCl_2$, then

$$[Cl] = K^{\frac{1}{2}} [Cl_2]^{\frac{1}{2}}$$

throughout the course of the reaction. The participation of Cl atoms as intermediates may therefore be guessed immediately. A second intermediate is required. If participation of ClCO is postulated, then a formation mechanism would be

$$Cl_2 \overset{K_1}{\rightleftarrows} 2Cl$$

$$Cl + CO + M \overset{K_2}{\rightleftarrows} ClCO + M$$

$$ClCO + Cl_2 \overset{k_3}{\rightarrow} Cl_2CO + Cl$$

The two equilibria imply

$$[ClCO] = K_2[Cl][CO] = K_2 K_1^{\frac{1}{2}} [Cl_2]^{\frac{1}{2}} [CO]$$

The rate of phosgene formation, limited by the third elementary reaction, is then

$$R_f = k_3[Cl_2][ClCO]$$
$$= k_3 K_2 K_1^{\frac{1}{2}} [Cl_2]^{\frac{3}{2}} [CO]$$

in agreement with experiment.

For the decomposition reaction the postulated mechanism assumes the same two equilibria as the formation reaction. The rate controlling step is the reverse of the rate controlling step in the formation reaction

$$COCl_2 + Cl \overset{k_4}{\rightarrow} Cl_2 + COCl$$

which implies the rate law

$$R = k_4 [COCl_2][Cl]$$
$$= k_4 K_1^{\frac{1}{2}} [COCl_2][Cl_2]^{\frac{1}{2}}$$

again in agreement with experiment.

Isomerization of cyclopropane to propylene. Cyclopropane (denoted Δ in the equations below) is a stable gas at room temperature, but it isomerizes irreversibly to propylene when heated.   It was shown that this reaction is not catalyzed by the walls of the reaction vessel or by added gases.   The rate law was found to be

$$R = k[\Delta]^{1.0}$$

at high pressures and

$$R = k'[\Delta]^{2.0}$$

at low pressures.   This kind of behavior is characteristic of gas-phase unimolecular (sometimes called " pseudo first order ") decompositions and isomerizations, which we consider in detail in Chapter 5.   For the present, we may obtain the observed rate law by postulating that a vibrationally energized cyclopropane Δ^* is an intermediate in the isomerization.   The bimolecular vibrational energization can be followed either by bimolecular deenergization or by unimolecular isomerization:

$$\Delta + \Delta \xrightarrow{k_1} \Delta^* + \Delta$$

$$\Delta^* + \Delta \xrightarrow{k_{-1}} \Delta + \Delta$$

$$\Delta^* \xrightarrow{k_2} CH{=}CH{-}CH_3$$

A steady state assumption applied to Δ^* gives

$$k_1[\Delta]^2 = k_{-1}[\Delta^*][\Delta] + k_2[\Delta^*]$$

$$[\Delta^*] = \frac{k_1[\Delta]^2}{k_2 + k_{-1}[\Delta]}$$

The isomerization rate is

$$R = k_2[\Delta^*] = \frac{k_2 k_1[\Delta]^2}{k_2 + k_{-1}[\Delta]}$$

High pressures of cyclopropane will increase the magnitude of $k_{-1}[\Delta]$ compared to k_2; the *high pressure limit* would prevail for $k_{-1}[\Delta] \gg k_2$ and

$$R = \frac{k_2 k_1}{k_{-1}} [\Delta]$$

At low pressures the inequality $k_1[\Delta] \ll k_2$ holds, in which case the isomerization rate becomes

$$R = k_1[\Delta]^2$$

Then the experimental rate constants k and k' are related to elementary reaction rate constants by

$$k = \frac{k_2 k_1}{k_{-1}} \quad \text{and} \quad k' = k_1$$

It turns out that many mechanisms give rate laws of the mass action form as limiting cases of more complicated but more general rate laws.

Detailed consideration of the unimolecular reaction mechanism is forthcoming in Chapter 5.

The Na + Cl₂ reaction. Formation of NaCl from the gaseous elements is one of the fastest gas reactions known. In fact, the conversion rate is faster than the encounter rate calculated for Na atoms and Cl_2 molecules using "normal" atomic and molecular diameters (Section 4-2). Such rapid reactions can be studied by allowing the two reactants to diffuse into one another at very low pressure. The conversion rate for the Na + Cl₂ reaction in particular was found from the thickness profile of the NaCl deposit formed on the wall of the tube in which the Na vapor and the Cl_2 gas were diffusing into one another. From the extremely high conversion rate it was concluded that the elementary reaction

$$Na + Cl_2 \rightarrow NaCl + Cl$$

must be the first step and account for the main sodium consumption. The Cl atom formed in this step will later form either Cl_2 or NaCl.

An interesting feature of the reaction is that yellow light, the so-called "D lines" of sodium atoms, is emitted. It is known that 48 kcal of energy are required to excite D line emission. Since the first step liberates only 41 kcal, it cannot provide the excitation. An elementary reaction that can is

$$Cl + Na_2 \rightarrow NaCl + Na$$

which is exothermic by 80 kcal. The participation of this reaction is indicated by the findings that the emission intensity is proportional to $[Na]^{2.0}$ and that the temperature dependence of the emission intensity is identical to that of the equilibrium $2Na \rightleftarrows Na_2$. Its exothermicity is sufficient to excite Na atoms to the electronic state responsible for the D line emission:

$$Na^\dagger \rightarrow Na + hv(\text{D lines})$$

It is not the only possible mechanism, however. Another, more indirect one, is

$$Cl + Na_2 \rightarrow NaCl^* + Na$$

$$NaCl^* + Na \rightarrow Na^\dagger + NaCl$$

$$NaCl^* + N_2 \rightarrow NaCl + N_2$$

$$Na^\dagger \rightarrow Na + hv(\text{D lines})$$

Transfer of energy from vibrationally excited $NaCl^*$ to Na is a slow reaction that can be suppressed by N_2 addition. This mechanism thus rationalizes all the observations.

For many years it was thought that only the indirect mechanism was correct. In 1971, however, a molecular beam experiment (Sec. 9.6) was reported in which a Cl atomic beam intercepted an Na_2 molecular beam. Observation of the D-lines emission from the intersection point of the beams proved that direct excitation occurs. The excitation by energy transfer from $NaCl^*$ was also shown to occur readily in another molecular beam experiment.

It is worth noting that deposition of energy of reaction in vibration of the newly formed bond in $NaCl^*$ is characteristic of atom transfer reactions. In many cases of exothermic reactions of the form

$$A + BC \rightarrow AB + C$$

or

$$A + BCD \rightarrow AB + CD$$

a substantial fraction of the energy released in the reaction is found in vibration of the A—B bond.

Debromination of stilbene dibromide by stannous chloride. Stilbene dibromide $(C_6H_5CHBr)_2$ in dimethylformamide solution is readily debrominated to stilbene $(C_6H_5CH)_2$ by $SnCl_2$. The rate law was found to be

$$R = k[(C_6H_5CHBr)_2]^{1.0} [SnCl_2]^{1.0}$$

From this rate law we can conclude that the rate limiting elementary reaction is either a bimolecular reaction of one molecule of each of the reactants or an unknown reaction of intermediates related by equilibria or steady states to the reactant concentrations in such a way that this second order rate law holds. We cannot decide between these possibilities on the basis of the rate law alone. It was further found, however, that if the *meso* optical isomer of stilbene dibromide was debrominated, only the *trans* geometrical isomer of stilbene was formed. When *dl* stilbene dibromide was debrominated, the product stilbene was found to 94% *trans* isomer and 6% *cis* isomer, with these percentages remaining constant over a substantial temperature range. These facts suggest that removal of the bromine atoms from the *meso* isomer occurs in a concerted step; that is, the olefinic bond is formed in such a way that rotation about the C—C bond during the reaction is impossible, whereas some rotation may occur when debrominating the *dl* compound. Taking the simpler *meso* case first, we can see from the projection formula that *trans*-stilbene can be formed only if the bromine atoms are opposite one another during the reaction

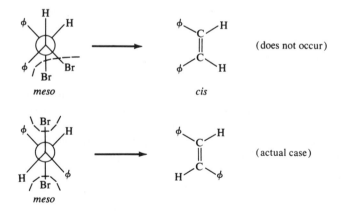

The bromine atoms must, therefore, depart in opposite directions during the debromination; the departure of the second atom must follow very quickly after the first atom departs so that reorientation around the C—C single bond does not occur. For the moment let us assume that the first bromine atom

leaves as Br^+, oxidizing $SnCl_2$ to $SnCl_2Br^+$ and leaving the carbanion

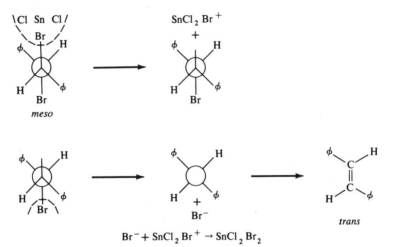

meso

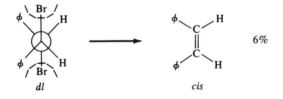

trans

$$Br^- + SnCl_2Br^+ \rightarrow SnCl_2Br_2$$

which immediately ejects a bromide ion to form *trans*-stilbene. Since we require that removal of the second atom be fast compared to a molecular rotation, the first step must be rate limiting.

For debromination of the *dl* compound the situation is not clear at all. A mechanism analogous to the *meso* case would yield the 6% *cis* isomer

dl   *cis*   6%

However, steric repulsion of the phenyl groups could cause rotation about the C—C bond of the carbanion to be more rapid than in the *meso* debromination and possibly more rapid than bromide ion loss. This repulsion would thus cause some fraction of the debromination from the conformation shown above to give the more stable *trans* isomer. Alternatively, other conformations of the *dl* isomer

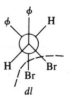

dl   *dl*

could also lose their bromine atoms simultaneously to give part of the 94%
trans isomer. Which of the possibilities is correct is difficult to say. The
observation that the percentage of *cis* isomer formed is independent of temperature suggests that perhaps neither of them is, but that some intermediate
species is produced in a displacement reaction.

The reducing agent attacking the stilbene dibromide molecules need not be
$SnCl_2$ but could be any ionic or solvated species whose concentration is related to the amount of $SnCl_2$ added by equilibria analogous to those mentioned in connection with the urea synthesis mechanism. Here also, we can
make no decision on kinetic or thermodynamic grounds as to what the
reducing species is.

3-8 ELEMENTARY REACTION RATE CONSTANTS

We saw in the examples introduced above that it is possible to utilize kinetic
measurements on reactions with simple mechanisms to obtain expressions
relating rate constants of experimental rate laws to rate constants (and equilibrium constants) of elementary steps involved in postulated mechanisms.
If the experimental results are correctly described by the mechanism, then it
may happen that one or more of the elementary reactions in the mechanism
may be assigned a rate constant. If the Arrhenius parameters of the experimental rate constant have been determined, then the Arrhenius parameters
of one or more of the elementary reactions concerned can be deduced.

This is a valuable scientific result, for starting from expressions for elementary reaction rate constants in the Arrhenius form we can construct
theories about how elementary reactions occur. The approaches used differ
according to whether the reactions are unimolecular, bimolecular, or termolecular and according to whether the reactions take place in the gas phase
or in solution. They have the common goal, however, of providing a
rationalization of all the rate constant expressions found experimentally for
a given class of elementary reactions. Insofar as possible, one tries to make
quantitative predictions of rate constants.

Theoretical interpretation of elementary reaction rates is a popular research
topic for theoretically minded chemists. In this book we cannot begin to
describe the entire subject. We devote the next two chapters to a consideration of the principal theoretical approaches that have met with success in
dealing with gas phase reactions, and Chapter 6 to the basic theories of solution reactions. We note here two important features of the theoretical work.
First, theoretical elementary reaction rate constants can be compared with
experiment only to the degree of reliability with which the mechanism of the
chemical reaction concerned is known. This means that when experimental
and theoretical elementary reaction rate constants are compared, allowance
must be made for error on both sides. This situation is quite different from

other comparisons of theory with experiment in physical chemistry, since outside of chemical kinetics one usually knows quite well what has been measured. Second, the theoretical work shows that it is indeed possible to make *a priori* estimates of Arrhenius parameters for elementary reactions, particularly for the ratios of the rate constants of related reactions. If rate studies are found to yield elementary reaction rate constant expressions that appear to be in substantial disagreement with theoretical values, the mechanism is suspect.

EXERCISES

3-3   Show that the time required for the concentration of A to decrease from its starting value $[A]_0$ to the value $f[A]_0$ by the reaction A → products is given by $t_f = -k^{-1} \ln f$ if the rate law is $-d[A]/dt = k_1[A]^{1.0}$ and $t_f = [A]^{1-n} (f^{1-n} - 1) \times (n-1)^{-1} k^{-1}$ if the rate law is $-d[A]/dt = k_n[A]^n, n \neq 1$. Show how n and k_n can be found from a log–log graph of experimental values for t_f and $[A]_0$.

3-4   Simultaneous formation of several products from one starting composition is a common occurrence in chemistry. The simplest example of this would be the parallel first order reactions A → B (1) and A → C (2), with the rate law $-d[A]/dt = k_1[A] + k_2[A]$. (a) Show that the fractional life of A is given by $t_f = -(k_1 + k_2) \ln f$. (b) What are the rate laws and integrated rate laws for [B] and [C] in terms of $[A]_0$ if $[B]_0 = [C]_0 = 0$? (c) Under what circumstances can the constants k_1 and k_2 be obtained independently by measuring the product ratio $[C]_\infty/[B]_\infty$ at long reaction times?

3-5   Most of the sun's energy is generated by the nuclear reaction mechanism

$$H + H \to D + e^+ + \nu \tag{1}$$

$$D + H \to {}^3He + \gamma \tag{2}$$

$$^3He + {}^3He \to {}^4He + 2H \tag{3}$$

$$e^- + e^+ \to 2\gamma \tag{4}$$

Where e^+ represents a positron, e^- an electron, ν a neutrino, and γ a gamma ray. (a) Show that the corresponding stoichiometric equation is $4H + 2e^- \to He + 2\nu + 6\gamma$. (b) Write the expression for R, the rate of this reaction, in terms of $d[He]/dt$ and $d[H]dt$. (c) The rate limiting step is the first elementary reaction. Show that application of the steady state approximation to the intermediates D and ^3He yields $[D] = k_1[H]/k_2$, $[^3He] = (k_2[D][H]/2k_3)^{\frac{1}{2}}$, and $d[He]/dt = k_1[H]^2/2$. (d) The sun radiates 3.9×10^{33} erg sec^{-1}. Evaluate the rate of reaction (1) assuming that all of the solar energy originates from this set of nuclear reactions. How many cubic kilometers of liquid hydrogen per second (at its earth density 0.07 g cm^{-3}) would be required to supply fuel for the sun?

3-6   In an experiment on the reaction $H_2 + Cl_2 \rightarrow 2HCl$ under irradiation by uv light the hydrogen pressure was found to decrease as follows:

t (min)	p_{H_2} (torr)
0	0.132
3	0.104
10.5	0.0606
20.5	0.0273
35.5	0.00715
55.0	0.00155

(a) Show that the reaction is approximately first order in hydrogen by constructing an appropriate graph.   (b) The reaction is also first order in chlorine.   In the experiment cited the initial chlorine pressure was 13.0 torr.   Calculate the second order rate constant in $torr^{-1}$ min^{-1} and liters $mole^{-1}$ sec^{-1}.

3-7   The rate at which the pesticide endrin is metabolized by rats was measured by injecting 200 μg of radioactive endrin per kilogram of rat weight.   The amount of endrin remaining in the rats was measured by radiochemical techniques as a function of time.   Determine the reaction order from the following data:

Days after injection:	0	1	2	3
^{14}C activity, male rats:	253	190	161	143
^{14}C activity, female rats:	196	137	102	82

4	5	6	7	8	9	10	15	20
124	110	98	88	82	73	66	49	45
71	63	55	49	43	37	34	23	15

The ^{14}C activity, given in arbitrary units, is proportional to the amount of endrin remaining in the rat.   Use a graphical method.

3-8   Pulse radiolysis (Sections 9-4 and 10-4) of cyclopentane produces a large concentration of cyclopentyl radicals in a short time, about 10^{-6} sec.   In pure cyclopentane the following data were obtained for the concentration of cyclopentyl radicals (given in arbitrary units) as a function of time (in microseconds) after the pulse.   Use a graphical method to find the reaction order for radical disappearance.

t	8	18	34	55	77	96	118	137
[R]	275	158	94	58	40	30	22	16

3-9   At 298, 308, and 323°K the second order rate constants for the alkaline hydrolysis of ethyl benzoate in 85% ethanol solvent were found to be 6.21, 16.8, and 62.8 $\times$ 10^{-4} liters $mole^{-1}$ sec^{-1}, respectively.   Find A and E_A for this reaction from a graph of log k versus $1/T$.

3-10   The following data were reported for the proton transfer reaction $NO_2C_6H_4CH_2CN + C_2H_5O^- \rightarrow NO_2C_6H_4CHCN^- + C_2H_5OH$:

Temperature °K	173	183	193	203
k liters $mole^{-1}$ sec^{-1}	3.4×10^1	1.5×10^2	5.0×10^2	1.8×10^3

Calculate A and E_A for this reaction.   (Answer: $A = 1.3 \times 10^{13}$ liters $mole^{-1}$ sec^{-1}, $E_A = 9.2$ kcal.)

3-11   Water has two nuclear spin isomers, *ortho* and *para* water, corresponding to two values of the nuclear spin quantum number $I = 1$ and $I = 0$.   The equilibrium constant for the interconversion reaction

$$A \underset{k_{-1}}{\overset{k_1}{\rightleftharpoons}} B$$

where A and B represent *para* and *ortho*, respectively, depends on temperature in such a way that $K = k_1/k_{-1} = 3.0$ at high temperatures and $K = 1.0$ at low temperatures.   When a crystal of argon containing a small amount of water was suddenly cooled from 30°K to 7°K, it was found by monitoring the ir absorption spectrum that establishment of the 7°K equilibrium was 50% complete in about 100 min. (a) Sketch a rough graph of mole fraction *ortho* water versus time.   Calculate k_1 assuming that the interconversion is a reversible first order process and that 30 and 7°K correspond to the high and low temperature conditions, respectively.   (b) When a small amount of O_2 was added to the crystal, the time required for 50% conversion decreased to 20 min.   If this effect is regarded as a decrease in effective activation energy for the interconversion, by what amount did the O_2 addition decrease the activation energy?

3-12   Electrical discharges in nitrogen produce an *afterglow*, which is light emission occurring in a complex mechanism that is found to be second order in [N].   The rate of photon production between 300°K and 2,400°K was reported to be

$$R = 1.1 \times 10^{-17} \, (T/300)^{-0.90} \, [N]^{2.0} \text{ photons cm}^{-3} \text{ sec}^{-1}$$

What is the second order rate constant in Arrhenius form?

3-13   The dissociative attachment reaction

$$e^- + NO^+ \rightarrow N + O$$

was found to have a rate constant $k(T) = 9 \pm 3 \times 10^{-7}$ cm³ sec⁻¹ at 196°K and $3.4 \pm 0.3 \times 10^{-7}$ cm³ sec⁻¹ at 358°K.   (a) Calculate E_A for this reaction.   (b) If $k(T)$ is written as $k(T) = BT^n$, what are B and n?   (c) At 3,000°K, $k(T)$ was found to be $4 \pm 2 \times 10^{-8}$ cm³ sec⁻¹.   Are both of your $k(T)$ functions consistent with this result?

3-14   A certain antibiotic A is metabolized with a rate law $-d[A]/dt = k_W(T)[A]^{1.0}$.   At human body temperature 98.6°F = 37.0°C = 310°K the first order rate constant k_W (310) is given by k_W (310) $= 3.0 \times 10^{-7}$ W sec⁻¹, where W is the body weight in pounds.   During treatment the antibiotic concentration should not decrease below 100 mg per 100 lb body weight.   (a) How frequently must a 170-lb man without fever take 500-mg antibiotic pills to prevent the antibiotic concentration from falling below this level?   (b) At 102.2°F = 312°K, it is found that k_W (312) = $4.0 \times 10^{-7} W$.   How frequently must a 170-lb man running a 102.2°F fever take 500-mg antibiotic pills to maintain the same minimum concentration?   What is the activation energy of the reaction that metabolizes the antibiotic?

3-15   The equilibrium constant for the reaction $Cl + H_2 \rightleftharpoons HCl + H$ is 0.305 at 298°K.   Check to see if the rate constant data in Table 4-1 confirm that $K_{eq} = k_f/k_r$ holds for this reaction.

3-16   Derive a justification for using the form $k(T) = BT^n \exp(-E_B/RT)$ by an argument analogous to that used in Section 2-5, but using for the ΔE term in $d \ln K_c/dT = \Delta E/RT^2$ a temperature-dependent form $E = E^0 + \int_0^T c\,dT$, where c is a constant.

3-17   The preexponential factors A and B (Section 3-4) are *approximately* related to one another by $A = B(e\,T)^m$, and E_A and E_B are approximately related by $E_A = E_B + mRT$.   (a) Derive these relationships by equating the corresponding derivatives $d\,k(T)/d(T)$ to one another.   (b) Sketch the $\log k(T)$ versus $1/T$ graphs of the two $k(T)$ functions that would be the best fits to the data: $k(333°K) = 1.0 \times 10^4$, $k(400°K) = 1.1 \times 10^5$, $k(500°K) = 1.0 \times 10^6$.   (c) If T in the above relationships is set equal to $400°K$, how would the Arrhenius equation graph using the values of E_A and A calculated from the best fit values of E_B, B, and m compare with the best fit Arrhenius equation graph? [*Hint:* What assumptions did you make in (a)?] (d) What value for $k(500°K)$ would be calculated from the second Arrhenius graph?   (e) Explain the mathematical fallacy of the above relationships.   Show with the aid of your graph that $E_A \simeq E_B + mRT$ is approximately true despite the fallacy.

3-18   In the $CO + O$ reaction (Section 3-7) it is found that $M = O_2$ gives emission intensity comparable to $M = N_2$.   This may be interpreted to mean that the rate-determining first step does not involve a change in spin multiplicity (Section 10-2) because elementary reactions that do involve spin multiplicity changes have higher rates when paramagnetic molecules such as O_2 are present.   Show by use of the steady state approximation that replacement of the rate-determining step by $CO + O + M \rightarrow CO_2^{\dagger\dagger} + M$ and addition of the step $CO_2^{\dagger\dagger} + M \rightarrow CO_2^{\dagger} + M$ does not change the rate law.   The designation $CO_2^{\dagger\dagger}$ indicates a $CO_2^{\dagger}$ molecule with the same spin multiplicity as $CO + O$, and $CO_2^{\dagger}$ indicates a molecule with the same spin multiplicity as CO_2.

3-19   (a) A second possible intermediate species for the phosgene formation mechanism (Section 3-7) is Cl_3. Show that a mechanism involving Cl_3 rather than $COCl$ as intermediate also gives the correct rate laws for formation and decomposition of $COCl_2$.

(b) Suppose that the formation of $COCl_2$ proceeds by way of $COCl$ but the decomposition proceeds by way of Cl_3.   Show that this supposition will not give the correct equilibrium constant when $R_d = R_f$, whereas either intermediate used consistently will.   The impossibility of such a "triangular" mechanism is called the *principle of detailed balancing*.   The mechanism for the forward reaction therefore implies the mechanism for the reverse reaction.

(c) Show how the activation energies for forward and reverse reactions could be used to derive the heat of formation of Cl_3 or $COCl$.   [Photochemical experiments (Chapter 10) indicate that $COCl$ is the more probable intermediate.]

SUPPLEMENTARY REFERENCES

Experimental methods for reaction rate measurements, mathematical techniques for treatment of rate data, and the use of rate data for discovering and testing mechanisms are thoroughly described in

Investigation of Rates and Mechanisms of Reactions, Vol. VIII, Parts I and II, of *Technique of Organic Chemistry*, S. L. Friess, E. S. Lewis, and A. Weissberger, Eds. (Interscience, New York, 1961 and 1963).

C. H. Bamford and C. F. H. Tipper, Eds., *Comprehensive Chemical Kinetics, Vol. 1: The Practice of Kinetics* (Elsevier Publishing Co., Amsterdam, 1969).

The relationships connecting thermodynamics and chemical kinetics are discussed in detail by

S. W. Benson, *Thermochemical Kinetics* (John Wiley & Sons, New York, 1968).

A systematic introduction to the use of rate laws in finding mechanisms is given by

J. O. Edwards, E. F. Greene, and J. Ross, "From Stoichiometry and Rate Law to Mechanism," in *J. Chem. Ed.* **45**, 381 (1968).

Chapter 4

BIMOLECULAR GAS REACTIONS

BIMOLECULAR gas reactions are the epitome of chemical change: two molecules encounter one another in free flight and suddenly lose their identity; out of the encounter emerge one, two, or perhaps three new chemical species. In this chapter we consider reactions in which two molecules are produced; the other possibilities are more appropriately considered as the reverses of unimolecular decompositions or of termolecular recombinations and will be discussed in Chapter 5. Many of the known bimolecular gas reactions involve simple species, that is, atoms and diatomic or triatomic molecules. In these cases there is an abundance of involvement between microscopic theories and experimental observations which promises that the atomic motions leading to chemical change in simple bimolecular gas reactions will eventually be known and understood in detail.

Our theoretical development will show that bimolecular rate constant formulas are not generally expected to be of the Arrhenius form. Experimental bimolecular rate constants are nevertheless usually reported in the Arrhenius form. A sampling of some experimental rate constants is given in Table 4-1. The variety of A and E_A values is seen to be great. It is the task of bimolecular reaction rate theory to correlate these experimental parameters with the properties of the atoms and molecules involved.

We shall develop the theory of bimolecular gas reactions by investigating the relationship of their rate constants to a quantity Q called the reactive cross section. Since the method we use in Section 4-1 to develop the necessary equations is somewhat lengthy, we give a preliminary introduction of the

TABLE 4-1 ARRHENIUS PARAMETERS AND STERIC FACTORS FOR SOME BIMOLECULAR GAS REACTIONS

Reaction	T range	$\log_{10} A$	E_A	p
$H + D_2 \rightarrow HD + D$	300–750	10.69	9.39	0.088
$D + H_2 \rightarrow HD + H$	250–750	10.64	7.61	0.094
$H + HCl \rightarrow H_2 + Cl$	200–500	10.36	3.50	0.039
$H + HBr \rightarrow H_2 + Br$	1,000–1,700	11.04	3.7	0.076
$H + HI \rightarrow H_2 + I$	667–800	10.70	0.7	0.037
$H + Cl_2 \rightarrow HCl + Cl$	273–5,200	11.0	5.3	0.074
$H + Br_2 \rightarrow HBr + Br$	1,000–1,700	11.97	3.7	0.19
$H + I_2 \rightarrow HI + I$	667–738	10.60	0.0	0.016
$H + C_2H_6 \rightarrow H_2 + C_2H_5$	304–1,500	11.12	9.71	0.10
$O + O_3 \rightarrow O_2 + O_2$	273–900	10.08	4.79	0.037
$O + OH \rightarrow O_2 + H$	300–2,200	10.48	0.0	0.14
$N + NO \rightarrow N_2 + O$	300–6,000	10.11	0	0.040
$N + O_2 \rightarrow NO + O$	180–5,000	9.92	7.1	0.034
$S + O_2 \rightarrow SO + O$	660–1,150	10.04	5.6	0.026
$Cl + H_2 \rightarrow HCl + H$	250–450	10.08	4.3	0.020
$Cl + ICl \rightarrow Cl_2 + I$	303–333	8.7	4.5	0.0015
$OH + CO \rightarrow CO_2 + H$	300–2,000	8.62	1.08	0.0014
$OH + H_2 \rightarrow H_2O + H$	300–1,200	10.36	5.15	0.046
$Na + CCl_4 \rightarrow NaCl + CCl_3$	520	11.35	(0)	0.37
$Na + HI \rightarrow NaI + H$	600	11.7	(0)	1.5
$K + HBr \rightarrow KBr + H$	600	11.6	(0)	1.2
$K + HCl \rightarrow KCl + H$	600	11.0	(0)	0.30
$CH_3 + C_6H_6 \rightarrow CH_4 + C_6H_5$	456–600	7.4	9.2	0.000027
$CH_3 + CCl_4 \rightarrow CH_3Cl + CCl_3$	363–418	10.4	12.9	0.03
$C_2H_5 + HI \rightarrow C_2H_6 + I$	536–576	8.92	1.10	0.0015
$CH_3O + CH_4 \rightarrow CH_3OH + CH_3$	398–523	8.8	11.0	0.00099
$CCl_3 + Cl_2 \rightarrow CCl_4 + Cl$	343–428	9.86	5.30	0.014
$BH_3 + BH_3CO \rightarrow CO + B_2H_6$	273–333	8.4	7.0	0.00029
$PH_3 + B_2H_6 \rightarrow PH_3BH_3 + BH_3$	249–273	6.5	11.4	0.0000074
$CO + O_2 \rightarrow CO_2 + O$	2,400–3,000	9.54	51.0	0.0043
$C \quad , \cdot N_2O \rightarrow CO_2 + N_2$	1,300–1,900	8.04	23.	0.00015
$CO + NO_2 \rightarrow CO_2 + NO$	498–746	8.7	27.8	0.0011
$F_2 + ClO_2 \rightarrow FClO_2 + F$	227–247	7.11	8.0	0.000052
$O_3 + NO \rightarrow NO_2 + O_2$	216–322	8.76	2.46	0.0020
$O^+ + O_2 \rightarrow O_2^+ + O$	185–576	9.80	−0.38	0.031 (0.012)
$O^+ + CO_2 \rightarrow O_2^+ + CO$	300–600	11.82	0.0	2.1 (1.0)
$O^- + NO_2 \rightarrow NO_2^- + O$	300	11.86	(0)	2.6 (1.0)
$O_3^- + NO_2 \rightarrow NO_3^- + O$	300	9.78	(0)	0.018 (.012)

Notes to Table 4-1: The units are T, °K; A, liter mole^{-1} sec^{-1}; and E_A, kcal. In some cases the tabulated results were derived from a single investigation, whereas other entries represent a composite result of several chemists' experiments. Activation energies denoted (0) are estimates of zero based on high values of p, The steric factors in parentheses are based on the Langevin rate constant expression $k = 2.3 \times 10^3 \, (\alpha/\mu)^{1/2} \, p \, \text{cm}^3 \, \text{sec}^{-1}$ (cf. Exercise 4-10).

underlying ideas here in order that the derivations which follow can be more easily understood.

Imagine a container filled with a gas mixture comprised of A atoms, all of which are moving with the same velocity w, and diatomic BC molecules, all of which are at rest. Let the number of A atoms per cubic centimeter be ^1N_A and the number of BC molecules per cubic centimeter be $^1N_{BC}$. We suppose that a bimolecular reaction

$$A + BC \rightarrow AB + C$$

will occur if and only if an A atom comes within the distance σ_R of the center of a BC molecule. In 1 sec an A atom will travel w cm through the gas. It will react during that second if there is a BC molecule within the volume $\pi\sigma_R^2 w$ that is the locus of centers of all BC molecules that could react with that A atom during the 1-sec interval. The probability that a particular A atom will react in a 1-sec interval is thus the product of the volume $\pi\sigma_R^2 w$ and the BC concentration $^1N_{BC}$. The number of reactions per cubic centimeter per second for all of the A atoms, that is, the reaction rate, is given by the product of the A concentration and the probability per second of an A atom undergoing reaction

$$R = \frac{d\,^1N_{AB}}{dt} = \pi\sigma_R^2 w\,^1N_A\,^1N_{BC}$$

$$= Q(w)w\,^1N_A\,^1N_{BC}$$

In this equation we have introduced the reactive cross section $Q(w)$ for the area $\pi\sigma_R^2$. By comparing this equation with our usual way of expressing bimolecular reaction rates we see that the rate constant for velocity w is related to the reactive cross section for velocity w by the equation

$$\boxed{k(w) = wQ(w)}$$

In a real reaction system the A atoms do not all move with the same velocity, nor do the BC molecules remain at rest. The motions of the A atoms and the BC molecules are instead both described by Maxwell–Boltzmann distribution functions. Since it is the relative velocity of A and BC that matters for the chemical reaction, we are primarily interested in the distribution function of relative velocity. Let us call the relative velocity u, the corresponding distribution function $f(T, u)$, and the corresponding reactive cross section $Q(u)$. We can then generalize the line of reasoning used in the previous paragraph to conclude that the probability per second that a particular

A atom reacts and has relative velocity in the range u to $u + du$ is $u\,Q(u)$ $f(T, u)\,{}^1N_{BC}\,du$. The total reaction rate is obtained by multiplying by the concentration of A atoms and integrating over the range of relative velocity

$$R = {}^1N_A\,{}^1N_{BC} \int_0^\infty Q(u)\,f(T, u)\,u\,du$$

This implies that the thermal average rate constant for the reaction is given by the equation

$$k(T) = \int_0^\infty Q(u)\,f(T, u)\,u\,du$$

The left-hand side of this equation is the elementary reaction rate constant $k(T)$ that one would obtain by measuring the reaction rate, whereas the right-hand side, containing the reactive cross section $Q(u)$, describes the reactivity of A and BC on the molecular scale.

4-1 CROSS SECTION FOR REACTIVE SCATTERING

Let us imagine that we are going to do experiments on a bimolecular chemical reaction in the same way that a physicist might do them on a nuclear reaction. The physicist uses a source of energetic particles, for instance 10-MeV protons, such as a cyclotron. The beam of protons from the cyclotron is directed at a target containing the element whose nuclear reactions with protons are to be investigated. Particle detectors measure the fraction of protons lost in the target, the energy lost by the protons in passing through the target, and the intensity of the proton or product particle flux as a function of angle away from the incident beam. All of these measurements taken together, and repeated for a range of incident proton energies, characterize the nuclear reactions under investigation. A schematic drawing of an experiment of this kind is shown in Figure 4-1.

Protons that leave the beam are referred to as *scattered* particles. Several different types of scattering can arise from interactions between the protons and the target nuclei. The simplest possibility is *elastic scattering*, which can be regarded as the nuclear analog of a billiard ball collision. When an incident proton transfers some of its energy to a target nucleus by raising the nucleus to one of its higher energy nuclear quantum states, *inelastic scattering* occurs. If a proton undergoes a nuclear reaction with the target nucleus, *reactive scattering* occurs. In the jargon of nuclear physics, the different possible types of scattering are called *channels*; there are *elastic channels*, *inelastic channels*, and *reactive channels*. The number and nature of the channels depends on the energy of the nuclear encounter and the nature of the particles involved. It is customary to regard each of the possible final quantum states for the scattering event as a different channel. Some of the possible final quantum states may be inaccessible for the conditions of a particular experiment, for example because the proton energy may be in the wrong range. Such inaccessible channels are called *closed channels*; those into which scattering is possible for particular experimental conditions are the *open channels*.

We are primarily interested in one particular concept that arises from experiments such as this: the concept of the *total reactive cross section*. The ratio

$$P_R(E) = \frac{(\text{number of protons reacting per second})}{(\text{number of protons striking target per second})}$$

can be regarded as the probability that an incident proton of energy E will undergo a nuclear reaction. Let us assume that the target is a thin one so that most of the protons pass through it without being scattered. From the point of view of a proton approaching the target, the above probability is then simply the chance that a nucleus with which the proton can react happens to lie in its path. If the area of the proton beam is $\mathscr{A}$ and the target has $N_{\mathscr{A}}$ atoms in the area exposed to the beam, then the above ratio gives the fraction of the beam area that is blocked by one of the $N_{\mathscr{A}}$ nuclei which can remove a

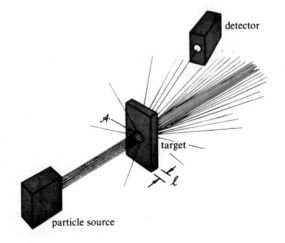

FIGURE 4-1 *Schematic drawing of a scattering experiment. The detector is usually movable in order that the intensity of scattering can be studied as a function of scattering angle. Sometimes detectors are used that can measure the energy of the scattered particles.*

proton by reaction. Since the blocking is done by the nuclei individually, we may assign an area

$$Q(E) \equiv \frac{P_R(E)\mathscr{A}}{N_{\mathscr{A}}}$$

to *each* of the nuclei in order to represent the equivalent area that a proton must hit in order to react with that nucleus; $Q(E)$ is called the *total reactive cross section* or, more simply, the *reactive cross section*. The functional dependence on the proton energy E indicates that reactive cross sections change with the energy of the incident beam particles.

The ideas introduced above can be carried over from nuclear physics to bimolecular gas reactions without change, although some additional ones are necessary also.   In fact, the same type of experiment can even be done in the laboratory, for favorable reactions, by the crossed molecular beam method (Section 9-6). We now examine the connection between the rate constant $k(T)$ and the reactive cross sections $Q(E)$ for a schematic bimolecular gas reaction between an atom A and a diatomic molecule BC

$$A + BC \rightarrow AB + C$$

First we must specify exactly what energy is meant by E.   Of several possibilities it is convenient to choose E to be the relative kinetic energy of A and the center of mass of BC.   Then we encounter a problem that is usually absent in considering nuclear reactions.   The structures of the molecules BC and AB are important in determining $Q(E)$.   Consequently, $Q(E)$ may depend on the rotational (J) and vibrational (v) quantum numbers of BC, and on the quantum numbers J' and v' of AB.   There are two ways to handle this complication.   One is to consider the BC quantum states separately and to express the reactive cross section as $Q(E, v, J)$, a function of E, v, and J; $Q(E, v, J)$ is thus the total cross section at relative kinetic energy E for production of AB in *any* of its rotational and vibrational quantum states starting from BC in the quantum state v, J.   An extension of this approach would be to consider the quantum states v', J' of AB separately as well; the cross section $Q(E, v, J, v', J')$ specifies the total cross section connecting channel v, J to channel v', J' for relative kinetic energy E.   This is a particularly appropriate procedure if one wishes to consider inelastic nonreactive scattering, that is, changes in v and/or J, as well as reactive scattering.   In this terminology we would divide the scattering channels into three groups: *elastic channels* where v, J, and E are unchanged and no reaction occurs; *inelastic channels* where v and/or J do change but no reaction occurs; and *reactive channels* where reaction occurs from v, J states of BC into v', J' states of AB.

For present purposes, however, a second way of handling the participation of BC quantum states is more suitable. We simply assume that Boltzmann distributions describe the fractions of BC molecules in the rotational and vibrational quantum states. If this is true, then a *thermal average* reactive cross section at temperature T can be found by a summation over the quantum states of BC

$$Q(E, T) \equiv \sum_{v=0}^{\infty} \sum_{J=0}^{\infty} f_v(T) f_J(T) \, Q(E, v, J)$$

In this equation $f_v(T)$, the fraction of BC molecules in vibrational state v, and $f_J(T)$, the fraction of BC molecules in rotational state J, are the Boltzmann distribution functions. From here on we consider reactive cross sections to be these averages $Q(E, T)$.

Before proceeding with our development we should emphasize that these considerations lead to meaningful $k(T)$ expressions only if the reacting system is, in fact, describable by Boltzmann distributions. This can indeed fail to be true in special circumstances, such as in hot atom reactions (Section 10-5) occurring in the upper atmosphere, where the relative energy E and the f_v and f_J distributions are more strongly influenced by solar radiation than by molecular collisions. For most conditions under which reactions are studied, however, there is good reason to believe that the assumption of a Boltzmann distribution is valid.

Let us note three further averages that have been included implicitly in $Q(E, T)$. First, a chemist intuitively senses that formation of AB is more likely to occur if A approaches the B end of BC rather than the C end; $Q(E, T)$ is thus an average over all possible orientations of BC at the time A approaches it. A similar average is taken over the phase of the vibrational motion of BC. Finally, the variables E, v, and J all refer to motions of the A, B, and C nuclei. It is assumed that the electrons move so rapidly during the scattering event that the three nuclei always move in an average potential (Section 4-4). This may be regarded as an extension of the Born–Oppenheimer approximation.

Two essential ingredients are required to relate $Q(E, T)$ to $k(T)$ for a given bimolecular gas reaction. One is the Boltzmann distribution function for E—which is known—and the other is the functional form of $Q(E, T)$ or $Q(E, v, J)$—which is a matter inspiring conjecture among theorists and ambitious experiments among practical chemists. For the remainder of this section we proceed on the assumption that $Q(E, T)$ is knowable, either theoretically or experimentally and derive an equation for $k(T)$. In later sections we deduce the consequences of some plausible theoretical assumptions about $Q(E, T)$ or $Q(E, v, J)$, we consider a selection of pertinent experimental results, and we extend the theory to more complicated cases than atoms and diatomic molecules.

The atoms A are now to be considered as the particles in the scattering experiment of Figure 4-1, and the target is a layer of gas, of thickness ℓ, composed of BC molecules at rest. The number of BC molecules in the beam path is the volume of target exposed $\mathscr{A}\ell$ times the molecular concentration $^1N_{BC}$. The total reactive area is $\mathscr{A}\ell\ ^1N_{BC}\ Q(E,\ T)$ and the reactive fraction of the beam area is $\ell\ ^1N_{BC}\ Q(E,\ T)$. The number of A atoms striking the target per second is $^1N_A\ w\mathscr{A}$, where ^1N_A is the number of A atoms per cubic centimeter of the beam and w is the velocity of the atoms. The number of reactive scattering events each second is $^1N_A\ ^1N_{BC}\ Q(E,\ T)w\mathscr{A}\ell$, and the number of reactive scattering events each second per unit volume of the target is $^1N_A\ ^1N_{BC}\ Q(E,\ T)w$.

Turning from the scattering experiment to an elementary reaction A + BC taking place in a mixture of gases A and BC, the number of A atoms per cubic centimeter moving with velocity between w and $w + dw$ is $f(w)\ ^1N_A\ dw$, where $f(w)$ is the velocity distribution function for A and the concentration of A is ^1N_A atoms cm^{-3}. The reaction rate for these A atoms would be $f(w)\ ^1N_A\ ^1N_{BC}\ Q(E,\ T)w\ dw$ by the above reasoning; the total conversion rate (in molecules $cm^{-3}\ sec^{-1}$) for all velocities would be the integral

$$R \equiv \frac{d\ ^1N_{AB}}{dt} = {}^1N_A\ ^1N_{BC} \int_0^\infty Q(E,\ T)f(w)w\ dw$$

Our line of argument so far contains one awkwardness and one flaw, which we now remove. The awkwardness is that we have written the reactive cross section as a function of E while the integral is over w; the flaw is that we have so far considered the BC molecules to be at rest while the A atoms approach at velocity w. In fact, the BC molecules also move and contribute significantly to the relative energy E. To obtain the correct form of the above integral we must convert to relative velocity, and to obtain a convenient form we transform further to an integral over relative energy.

Since we assume that the velocities of A and BC are described by Maxwell–Boltzmann distributions, the number of A atoms with velocity components in the range u_{Ax} to $u_{Ax} + du_{Ax}$, u_{Ay} to $u_{Ay} + du_{Ay}$, u_{Az} to $u_{Az} + du_{Az}$ is given by

$$^1N_A\ f(u_{Ax},\ u_{Ay},\ u_{Az})\ du_{Ax}\ du_{Ay}\ du_{Az}$$

$$= {}^1N_A \left[\frac{m_A}{2\pi kT}\right]^{\frac{3}{2}} \exp[-m_A(u_{Ax}^2 + u_{Ay}^2 + u_{Az}^2)/2kT]\ du_{Ax}\ du_{Ay}\ du_{Az}$$

where m_A is the mass of A. A similar distribution describes the velocities of the BC molecules. It is convenient to describe the location of an A atom by the vector $\mathbf{r}_A$, whose time derivative is the A velocity vector $\mathbf{u}_A$, and the location of a BC molecule by the vector $\mathbf{r}_{BC}$, whose time derivative is the BC velocity vector $\mathbf{u}_{BC}$. The squared magnitudes of the two velocity vectors, $u_A{}^2 \equiv \mathbf{u}_A \cdot \mathbf{u}_A$ and $u_{BC}^2 \equiv \mathbf{u}_{BC} \cdot \mathbf{u}_{BC}$, multiplied by $m_A/2$ and $m_{BC}/2$, respectively, give the kinetic energies of A and BC. A transformation to center-of-mass (c.m.) coordinates is made by introducing the variables

$$M \equiv m_A + m_{BC}$$

$$\mu \equiv \frac{m_A m_{BC}}{M}$$

$$\mathbf{R} \equiv \frac{m_A}{M} \mathbf{r}_A + \frac{m_{BC}}{M} \mathbf{r}_{BC}$$

$$\mathbf{U} \equiv \frac{m_A}{M} \mathbf{u}_A + \frac{m_{BC}}{M} \mathbf{u}_{BC}$$

$$\mathbf{r} \equiv \mathbf{r}_A - \mathbf{r}_{BC}$$

$$\mathbf{u} \equiv \mathbf{u}_A - \mathbf{u}_{BC}$$

$$u \equiv [\mathbf{u} \cdot \mathbf{u}]^{\frac{1}{2}}$$

In the new velocity variables the product

$$du_{Ax}\, du_{Ay}\, du_{Az}\, du_{BCx}\, du_{BCy}\, du_{BCz}\, {}^1N_A\, f(u_{Ax}, u_{Ay}, u_{Az})$$

$$\times\ {}^1N_{BC}\, f(u_{BCx}, u_{BCy}, u_{BCz})$$

becomes

$${}^1N_A\, {}^1N_{BC} f(\mathbf{U})\, f(\mathbf{u})\, du_x\, du_y\, du_z\, dU_x\, dU_y\, DU_z$$

where

$$f(\mathbf{U})f(\mathbf{u}) = \frac{\exp[-M\mathbf{U} \cdot \mathbf{U}/2kT - \mu\mathbf{u} \cdot \mathbf{u}/2kT]}{(2\pi kT/M)^{\frac{3}{2}}(2\pi kT/\mu)^{\frac{3}{2}}}$$

We recognize here that

$$\mu\mathbf{u} \cdot \mathbf{u}/2 = \mu[(u_{Ax} - u_{BCx})^2 + (u_{Ay} - u_{BCy})^2 + (u_{Az} - u_{BCz})^2]/2 = \mu u^2/2$$

is the relative kinetic energy E.

Our original equation for the reaction rate now becomes

$$R = \frac{{}^1N_A\,{}^1N_{BC}}{(2\pi kT/M)^{\frac{3}{2}}(2\pi kT/\mu)^{\frac{3}{2}}} \int\!\!\int\!\!\int\!\!\int\!\!\int\!\!\int_{-\infty}^{+\infty} \exp[-M\mathbf{U}\cdot\mathbf{U}/2kT - E/kT]$$

$$\times\; Q(E,\,T)\; u\; du_x\; du_y\; du_z\; dU_x\; dU_y\; dU_z$$

Integration over U_x, U_y, and U_z yields $(2\pi kT/M)^{\frac{3}{2}}$, which cancels with the same factor in the denominator; converting $du_x\,du_y\,du_z$ to $u^2 \sin\theta\;d\theta\;d\phi\;du$ and integrating over θ and ϕ yields 4π, after which the reaction rate expression becomes

$$R = \frac{4\pi\,{}^1N_A\,{}^1N_{BC}}{(2\pi kT/\mu)^{\frac{3}{2}}} \int_0^\infty e^{-E/kT}\,Q(E,\,T)\,u^3\,du$$

This removes the flaw in our earlier result; to remove the awkwardness that the cross section is a function of E while the integration is over u, we use $\mu u\,du = dE$ and $\mu u^2/2 = E$ to transform to an integral over E:

$$R = [8/\pi\mu(kT)^3]^{\frac{1}{2}}\,{}^1N_A\,{}^1N_{BC} \int_0^\infty e^{-E/kT}\,Q(E,\,T)\,E\,dE$$

By comparison with the second order conversion rate expression in cm^3 sec^{-1}, $R = k(T)\,{}^1N_A\,{}^1N_{BC}$, we see that the rate constant for the bimolecular elementary gas reaction $A + BC \rightarrow AB + C$ is

$$k(T) = [8/\pi\mu(kT)^3]^{\frac{1}{2}} \int_0^\infty e^{-E/kT}\,Q(E,\,T)\,E\,dE$$

in $cm^3\;sec^{-1}$. *This equation should be recognized as a fundamental link between the macroscopic and microscopic aspects of bimolecular reactions. The left-hand side is a quantity that determines the macroscopic conversion rate, while the right-hand side describes the interaction between* A *and* BC *at the molecular level.*

We have noted before that in deriving the fundamental rate constant equation it is assumed that both the relative motion and the internal quantum states are described by Boltzmann distributions, and that there are good reasons to believe that this is a valid assumption for bimolecular gas reactions under most conditions. One can see by inspection that an entirely different equation will result if the relative motion is non-Boltzmann, whereas if only the internal quantum state distributions are non-Boltzmann, the same integral, with $Q(E)$ averaged over whatever internal quantum state distribution pertains, can be used.

Since it was not necessary in the derivation to make use of the fact that A represented an atom and BC a diatomic molecule, the fundmental rate constant equation is valid for any bimolecular reaction.

If we presume that the reactive cross section is known for all quantum states of A and BC, then the rate formula could be written as a double sum of integrals

$$R = [8/\pi\mu(kT)^3]^{\frac{1}{2}} \sum_i \sum_j {}^1N_{A_i} {}^1N_{BC_j} \int_0^\infty Q(E, i, j)\, e^{-E/kT}\, E\, dE$$

where i and j index the quantum states of A and BC, respectively. If those quantum states are populated in a Boltzmann distribution, then the fundamental rate constant expression can be written

$$k(T) = [8/\pi\mu(kT)^3]^{\frac{1}{2}} q_A^{-1} q_{BC}^{-1} \sum_i \sum_j e^{-(\varepsilon_i + \varepsilon_j)/kT} \int_0^\infty Q(E, i, j)\, e^{-E/kT}\, E\, dE$$

where q_A and q_{BC} are the partition functions for the internal quantum states of A and BC.

In Sections 4-4 and 4-5 we consider theories and experiments that provide information about $Q(E, T)$. Our first use of the $k(T)$ equation, however, will be to evaluate the integral over E using simple assumed functional forms for $Q(E, T)$.

EXERCISE 4-1 (a) Show that the kinetic energy expression for A and BC in c.m. coordinates is equivalent to the kinetic energy expression for A and BC in laboratory coordinates. (b) Show that the phase integral over $\exp[-MU^2/2kT - \mu u^2/2kT]$ is equal to the phase integral over $\exp[-m_A u_A^2/2kT - m_{BC} u_{BC}^2/2kT]$. (c) Write out in detail the algebra leading from the transformation into c.m. coordinates to the $k(T)$ equation. (d) The factor $[8/\pi\mu(kT)^3]^{\frac{1}{2}}$ may also be written $[8kT/\pi\mu]^{\frac{1}{2}}(kT)^{-2}$. What is the significance of this? (*Hint*: Suppose $Q(E,T)$ is a constant.)

4-2 COLLISION THEORY OF BIMOLECULAR REACTION RATES

To derive an explicit rate constant equation we must evaluate the integral over E; to evaluate the integral over E we must know the function $Q(E, T)$. Once the integral has been worked out, we have an expression for $k(T)$ that is customarily called a *collision theory rate constant*.

A simple starting assumption would be a cross section that is the same for all internal quantum states and for all relative energies greater than a *threshold energy* E_0

$$Q(E, T) = 0 \qquad E < E_0$$

$$Q(E, T) = \pi\sigma_R^2 \qquad E > E_0$$

where $\sigma_R = r_A + r_{BC}$ (cf. Figure 4-2) is a reactive collision diameter analogous to the hard sphere collision diameter σ of the kinetic theory of gases. (Note that r_A and r_{BC} represent the radii of A and BC, not the magnitudes of the location vectors $\mathbf{r}_A$ and $\mathbf{r}_{BC}$ used in Section 4-1.) For this "*hard sphere cross section*"

$$k(T) = [8/\pi\mu(kT)^3]^{\frac{1}{2}} \int_0^\infty e^{-E/kT} Q(E, T)E \, dE$$

$$= [8/\pi\mu(kT)^3]^{\frac{1}{2}} \int_{E_0}^\infty e^{-E/kT} \pi\sigma_R^2 E \, dE$$

$$= [8/\pi\mu(kT)^3]^{\frac{1}{2}} \pi\sigma_R^2 (kT)^2 \, e^{-E_0/kT}\left(1 + \frac{E_0}{kT}\right)$$

$$= \left(\frac{8\pi kT}{\mu}\right)^{\frac{1}{2}} \sigma_R^2 e^{-E_0/kT}\left(1 + \frac{E_0}{kT}\right)$$

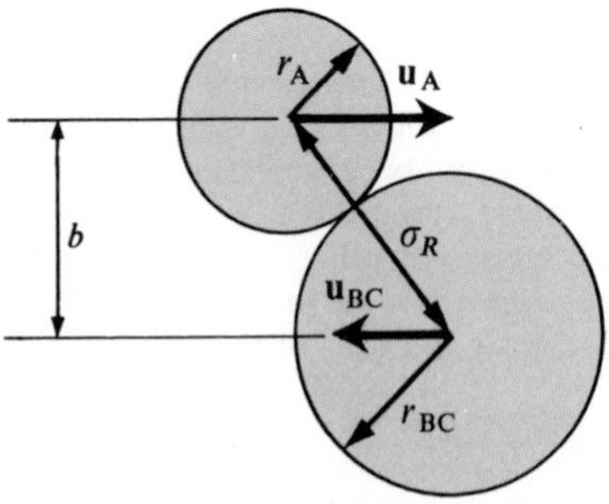

FIGURE 4-2 *Coordinate system used to describe collisions between atoms or molecules. In the collision shown here, atom A and molecule BC, assumed to be spheres of radii r_A and r_{BC}, collide with impact parameter b and relative velocity vector $\mathbf{u} \equiv \mathbf{u}_A - \mathbf{u}_{BC}$. The use of spheres is a convenient fiction; an actual collision would be more properly described by the forces on A, B, and C throughout the collision (cf. Section 4-4).*

It does not appear to be reasonable, however, that E_0 should be the same for all values of the impact parameter b (Figure 4-2); for a given relative velocity vector $\mathbf{u}$, a head-on collision ought to have a greater chance of leading to reaction than a grazing one. A plausible way of incorporating this idea into a $Q(E, T)$ function is to assume first that b must be less than a value σ_R for $Q(E, T)$ to be nonzero, and second that only the component of $\mathbf{u}$ along the line connecting the centers of A and BC is effective in promoting reaction. For each value of $\mathbf{u}$ such that $\mathbf{u} \cdot \mathbf{u}/2\mu > E_0$, there will be a maximum value of the impact parameter, b_0, at which the line-of-centers component, given by $w = u \cos \phi$, where ϕ is the angle between $\mathbf{u}$ and the line of centers at the time of contact, satisfies the equation $\mu w^2/2 = E_0$. For $b < b_0$, $\mu w^2/2$ is greater than E_0. From geometry, if $b < \sigma_R$, ϕ is given by $\sin \phi = b/\sigma_R$. Thus $w_0^2 = 2E_0/\mu$ implies a critical angle ϕ_0 through $u^2 \cos^2 \phi_0 = 2E_0/\mu$. Using the trigonometric identity $\sin^2 \phi_0 + \cos^2 \phi_0 = 1$ we have $u^2(1 - \sin^2 \phi_0) = 2E_0/\mu$, or $u^2(1 - b_0^2/\sigma_R^2) = 2E_0/\mu$, which with $E = \mu u^2/2$ rearranges to

$$
\begin{array}{ll}
b_0^2 = \sigma_R^2(1 - E_0/E) & E \geq E_0 \\
Q(E, T) \equiv \pi b_0^2 = \pi \sigma_R^2(1 - E_0/E) & E \geq E_0 \\
= 0 & E < E_0
\end{array}
$$

This $Q(E, T)$ function is called the *line-of-centers cross section*. When it is used in the fundamental rate constant equation, the $k(T)$ expression becomes

$$
k(T) = \left(\frac{8\pi kT}{\mu}\right)^{\frac{1}{2}} \sigma_R^2 e^{-E_0/kT}
$$

This particular rate constant equation is often called "*the*" *collision theory rate constant* equation. We shall refer to it hereafter as the *line-of-centers rate constant*.

(An important modification of the line-of-centers rate constant formula is introduced later in this section.)

EXERCISE 4-2 (a) Carry out the integration required to show that the line-of-centers cross section gives the above collision theory rate constant equation. (b) How is E_A related to E_0 for this cross section? [*Note*: E_A is a *molar quantity* and appears in Boltzmann factors as $\exp(-E_A/RT)$; E_0 is a *molecular quantity* appearing in Boltzmann factors as $\exp(-E_0/kT)$.] Consider R to be given in cal mole^{-1} deg^{-1} and k to be given in erg deg^{-1}.

The meaning of this equation becomes clear if we set E_0 equal to zero and replace the reactive collision diameter σ_R with the hard sphere collision diameter σ of the kinetic theory of gases. Then the right-hand side becomes

$(8\pi kT/\mu)^{\frac{1}{2}}\sigma^2$, an expression known as the *hard sphere collision frequency* (or *collision number*) and usually denoted Z. (Note that Z has the dimensions of a second order rate constant and is neither a number nor a frequency.) If we desire to calculate the total number of encounters per second between A atoms and BC molecules in a 1-cm^3 volume containing ^1N_A A atoms and $^1N_{BC}$ BC molecules, assuming A and BC to be hard spheres, we would compute the *encounter rate* from

$$R_E = \left(\frac{8\pi kT}{\mu}\right)^{\frac{1}{2}}\sigma^2 \, ^1N_A \, ^1N_{BC} = Z \, ^1N_A \, ^1N_{BC}$$

If instead we desire to calculate the *reaction rate* (in molecules cm^{-3} sec^{-1}) using the line-of-centers rate constant, we would write

$$R_{A+BC} = \left(\frac{8\pi kT}{\mu}\right)^{\frac{1}{2}}\sigma_R^2 e^{-E_0/kT} \, ^1N_A \, ^1N_{BC}$$

The line-of-centers reaction rate expression thus differs from the encounter rate expression in two respects: (1) σ_R replaces σ; and (2) a Boltzmann factor in the threshold energy $\exp(-E_0/kT)$ is included. In the framework of the collision theory of bimolecular reaction rates, then, the difference between a hard sphere collision and a reactive encounter is both geometrical—σ_R is not equal to σ—and energetic—requiring the line-of-centers component of E to be greater than E_0 for reaction adds a Boltzmann factor to the equation.

We now wish to compare line-of-centers rate constants with experimental values of $k(T)$ for bimolecular reactions.

Since there is as yet no adequate theoretical way to calculate E_0, we have to estimate it (assuming our theoretical model to be correct!) from the experimental $E_A \equiv -R[d \ln k(T)/d(1/T)]$ cal. As derived above [Exercise 4-2(b)], the required relationship is $E_0 = 6.95 \times 10^{-17} (E_A - RT/2)$ ergs.

Comparing the Arrhenius rate constant formula with the line-of-centers rate constant formula shows that the Arrhenius A factor corresponds to a factor

$$A = e^{\frac{1}{2}}\left(\frac{8\pi kT}{\mu}\right)^{\frac{1}{2}}\sigma_R^2$$

in the line-of-centers rate constant formula, where $e^{\frac{1}{2}}$ results from the conversion of the exponential factor from $\exp(-E_A/RT)$ to $\exp(-E_0/kT)$. If σ_R^2 were known, A could be calculated; but σ_R^2 is not known. To permit a comparison between theory and experiment, we define the ratio p

$$p \equiv \frac{\pi\sigma_R^2}{\pi\sigma^2}$$

where $\pi\sigma_R{}^2$ is divided by the kinetic theory of gases hard sphere cross section $\pi\sigma^2$. The collision diameter σ is presumably known experimentally from transport property measurements. This gives

$$A = pe^{\frac{1}{2}}\left(\frac{8\pi kT}{\mu}\right)^{\frac{1}{2}}\sigma^2$$

The factor p now expresses the correlation between theory and experiment. If $p \approx 1$, we conclude that the line-of-centers rate constant formula implies that the reactive cross section is about equal to the kinetic theory of gases cross section; if $p \ll 1$, we conclude that the reactive cross section is much smaller than the kinetic theory cross section; for the few cases where $p > 1$, we conclude that the reactive cross section is greater than the kinetic theory cross section.

In Table 4-1 are listed the parameters for a number of representative bimolecular gas reactions. It can be seen that the values of p cover quite a large range. The general trend is that p decreases with increasing reactant complexity. For this reason p has come to be called a *steric factor*. It expresses the idea that not all relative orientations of reactants during an encounter will lead to reaction. The more complex the reaction is, the more stringent the orientation requirement must be, and the smaller will be the value of p.

Unfortunately, it is no more possible to compute p from molecular parameters than it is to compute E_0. This is the great shortcoming of the collision theory of reaction rates: the theory is not an operational, quantitative one, nor can it become so until $Q(E, T)$ functions are known experimentally. It appeals to chemists because the results, such as the line-of-centers rate constant formula with $\sigma_R{}^2$ replaced by $p\sigma^2$

$$k(T) = p\left(\frac{8\pi kT}{\mu}\right)^{\frac{1}{2}}\sigma^2 e^{-E_0/kT} = pZe^{-E_0/kT}$$

are transparent, plausible connections between molecular theory and observed reaction rates.

Since p is found to be less than unity for all reactions except those of alkali metal atoms, we can use the above formula to compute *upper limits* to the rates of bimolecular gas reactions. The upper limit (excepting the special cases of alkali metal reactions) is obtained by setting p equal to unity and E_0 equal to the smallest value permitted by the thermochemistry of the elementary reaction. For an exothermic elementary reaction the smallest value of the threshold energy is $E_0 = 0$; for an endothermic elementary reaction the smallest value is the heat of reaction $E_0 = \Delta H_T^0$. A simple physical idea

emerges from our formula: the fastest rate possible for a bimolecular gas reaction is determined by the total number of collisions per second with sufficient energy to allow reaction to occur.   If internal degrees of freedom contribute to E_0, however, the activation-in-many-degrees-of-freedom rate
constant (cf. Section 4-3) will give a larger upper limit than the line-of-centers
rate constant.

An important modification to the line-of-centers rate constant formula
arises in ion–molecule reactions.   The strongly attractive electrostatic force
leads to a reactive cross section proportional to u^{-1} and an upper limit rate
constant formula $k = 2 \times 10^3 \, (\alpha/\mu)^{\frac{1}{2}} \, cm^3 \, sec^{-1}$ that is independent of temperature.   In this formula α is the polarizability of the molecule (cf. Exercise
4–10).

It is an unfortunate fact that quantum mechanical theories are not yet far
enough developed to be able to provide p or E_0.   This prevents us from
predicting $k(T)$ functions by use of the equations derived in this section, but
it does not prevent chemists from using the ideas developed here as a framework for interpreting experimental $k(T)$ data for bimolecular gas reactions,
and this is in fact customary practice.   The detailed attention we have given
to the ideas of this section is justified by the appropriateness of these ideas
for interpretive purposes rather than by the usefulness of the equations for
computational purposes.

It is necessary and desirable that $k(T)$ equations usable for computations
be developed also.   In Section 4-7 we present an alternative theoretical
framework that does provide such equations.

EXERCISE 4-3   (a) Show that a computational formula for p is $p = 2.2 \times 10^{-10}$
$\times A\mu^{\frac{1}{2}} \, T^{-\frac{1}{2}} \, \sigma^{-2}$ if A is in liters mole^{-1} sec^{-1}, μ is in atomic mass units, and σ is in
angstroms.   (b) Verify one of the p values listed in Table 4-1.   Since experimental
A values are derived from functions of *inverse* temperature, the temperature to use
in the computational formula is the average value of inverse temperature $2 \, T_1 T_2$
$\times (T_1 + T_2)^{-1}$, where $T_1 < T < T_2$ is the temperature range of the experiments.
If you cannot find σ values in a table, use your knowledge of atomic sizes and
bond lengths to estimate σ.

EXERCISE 4-4   The rate constant for the reaction $O + N_2^+ \rightarrow NO^+ + N$ was
found to be $2.5 \times 10^{-10} \, cm^3 \, sec^{-1}$ at 300°K.   Calculate the line-of-centers limiting
cross section $\pi\sigma_R^2$ assuming $E_0 = 0$.

EXERCISE 4-5   The reactive cross section for the reaction $NO_2 + M \rightarrow NO$
$+ OM$, where M represents a μ meson, was found to be $23 \times 10^{-16} \, cm^2$.   Compute the line-of-centers rate constant for this reaction at 300°K assuming $E_0 = 0$.

4-3   REFINEMENTS OF THE COLLISION THEORY

The two simple cross section functions assumed in the previous section have
a major flaw in that $Q(E, T)$ remains constant at very large E and a minor
flaw in that the cross section rises very steeply from zero at $E = E_0$, neither

of which is likely to be true for any real reaction. If it is assumed that the fragmentary information about cross sections now available from crossed molecular beam experiments, including ion and electron scattering experiments, and hot-atom experiments is characteristic of reactive cross sections for all bimolecular gas reactions, then the truth about the dependence of reactive cross sections on E is quite different from our previous assumptions. At $E = E_0$ the cross section actually rises fairly rapidly to a peak value and then gradually decreases, eventually decaying to zero. Near the peak value of the cross section graph a considerable amount of structure can occur that manifests the vibrational and rotational energy levels of the molecules involved. A sketch of a likely cross section graph for a reaction of an atom with a diatomic molecule is shown in Figure 4-3.

For the interpretation of reaction kinetics, the initial rise and the eventual decay of $Q(E, T)$ are more important than finer variations attributed to quantum effects. For lack of experimental data on the correct functions to use for an approximate description of $Q(E, T)$, we have to invent functions that possess the right general behavior near E_0 and for large E. One such function is

$$Q(E, T) = 0 \qquad E \leq E_0$$

$$Q(E, T) = \alpha(E - E_0)^n \exp[-\beta(E - E_0)] \qquad E > E_0$$

This function for $Q(E, T)$ leads to

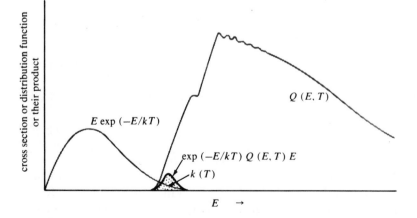

FIGURE 4-3 *Comparison of relative energy distribution function and cross section. The k(T) integral can be seen to have small contributions from energy much greater than the threshold energy E_0.*

$$k(T) = [8/\pi\mu(kT)^3]^{\frac{1}{2}}\left[\frac{\alpha\Gamma(n+1)}{c^{n+1}}\right]\left[\frac{n+1}{c} + E_0\right]e^{-E_0/kT}$$

$$c = \beta + 1/kT$$

where $\Gamma(n + 1)$ is the gamma function; $\Gamma(n + 1) = n!$ for positive integral n. The principal temperature dependence is again exponential, just as in the line-of-centers rate constant and in the Arrhenius equation. Unless some information about the parameters in the realistic cross section function were available from other sources, this rate constant function would not be distinguishable experimentally from the rate constant functions derived in the previous section.

EXERCISE 4-6   (a) Derive the above rate constant function.   (b) Show that $\beta = n/(E_{max} - E_0)$, where E_{max} is the value of E where $Q(E,T)$ has its maximum value. (c) For the case that $E_0 \gg kT$ and $(E_{max} - E_0) \gg nkT$, show that $E_A = E_0 + (n - \frac{1}{2})kT$.

In the distant future, chemists might be able to measure not only $Q(E, T)$ but even $Q(E, v, J)$, the $Q(E)$ function for individual internal quantum states of reactant molecules. From these measurements the summation method mentioned earlier would yield a $k(T)$ function.   There is already strong indirect evidence that $Q(E)$ differs greatly for different internal quantum states in some elementary reactions. For one example, the isotopic exchange reaction

$$^{15}N^{15}N + {}^{14}N^{14}N \rightarrow {}^{14}N^{15}N + {}^{14}N^{15}N$$

has been found to have small cross section for $v < 14$, but large cross section for $v > 14$. For another, the cross sections for the reaction

$$H + Cl_2 \rightarrow HCl + Cl$$

vary greatly for the several vibrational states of HCl that can be populated from the heat liberated in the reaction; although the $v = 0$ channel is open as far as the reaction energetics are concerned, the experimental cross section for producing HCl in the $v = 0$ state is zero.   Calculating rate constants from cross section functions for different quantum states is an appealing goal for future chemical research.   At present, however, even for the simplest case of A + BC, there are neither theories nor experiments that furnish the values of $Q(E)$ for successive vibrational and rotational states of BC with enough accuracy to justify such a calculation.   In view of the scarcity of $Q(E)$ information, we might wish to use the fundamental rate constant equation in reverse, as an integral equation for $Q(E)$, and derive cross section functions from rate measurements.   This unfortunately appears to be unachievable in practice, because it would require extremely precise $k(T)$ measurements to define the preexponential temperature dependence of $k(T)$ (cf. Section 3-4). Even so, only the values of $Q(E)$ just above the threshold energy would be derived from such measurements, a consideration to which we return in Section 4-5.

In summary, the situation with regard to using realistic cross section functions in the fundamental rate constant equation is that although such functions are much better representations of nature than the ones used in the previous section, they lead to rate constant functions that cannot be distinguished experimentally from the simple line-of-centers rate constant function.

We turn now from considering the consequences of using realistic cross section functions to take up a classical theory that considers the consequences of assuming that other degrees of freedom besides the relative motion play a major role in the reaction. There is a straightforward classical approach to formulating a $k(T)$ expression that takes internal degrees of freedom into account. This approach, called the *activation-in-many-degrees-of-freedom* assumption, is based on the simple idea that some or all of the internal degrees of freedom of the reactant molecules can contribute to the threshold energy E_0. The rate constant $k(T)$ is assumed to be the product of the hard sphere collision frequency, which is known from the kinetic theory of gases, and the probability that the total energy in a specified number of degrees of freedom of the colliding molecules exceeds E_0. In order to derive an explicit equation for $k(T)$, it is assumed that classical formulas for the average, thermal equilibrium energy contained in internal degrees of freedom apply. For mathematical convenience all of the participating degrees of freedom are regarded as vibrational; that is, we consider the energy to be stored in a collection of oscillators. If, using the methods of statistical mechanics and assuming that E_0 is much greater than kT, we compute the fraction of the time that a collection of s classical oscillators has total energy in excess of the value E_0, one obtains the expression

$$\left(\frac{E_0}{kT}\right)^{s-1} e^{-E_0/kT}\, \Gamma(s)^{-1}$$

This time fraction is set equal to the fraction of kinetic theory hard sphere collisions in which s "effective oscillators" have total energy in excess of the threshold energy for reaction E_0. This leads to the rate constant expression (in $cm^3\ sec^{-1}$)

$$\boxed{k(T) = \left(\frac{8\pi kT}{\mu}\right)^{\frac{1}{2}} \sigma^2 \Gamma(s)^{-1} \left(\frac{E_0}{kT}\right)^{s-1} e^{-E_0/kT}}$$

where the factor $(8\pi kT/\mu)^{\frac{1}{2}}\, \sigma^2 = Z$ is again the hard sphere collision frequency for collision diameter σ.

This equation, called the rate constant for *activation in many degrees of freedom*, has been applied to numerous problems of bimolecular reaction kinetics in which participation of internal degrees of freedom is thought to be important.   We shall consider two of them in the next chapter.   A complete derivation of the above equation, although not difficult, is regrettably lengthy and will not be given here.

For $s = 1$ the above equation reduces to the line-of-centers rate constant derived in the previous section. The two degrees of freedom concerned are translational, namely, the components of the motions of A and BC along the line joining their centers.   Since the inequalities $E_0 \gg kT$ and $(E_0/kT)^s \gg \Gamma(s)$ hold for the temperature ranges of most rate measurements, contributions of additional internal degrees of freedom act to *increase* $k(T)$, in accordance with what we would intuitively expect.   The maximum value of s for a reaction involving a total of N atoms is $3N - 3$, where subtracting 3 accounts for the c.m. motion.

In using the activation-in-many-degrees-of-freedom rate constant one should keep in mind that it stems from an explicit supposition about the dependence of the cross section on internal degrees of freedom (and also assumes validity of the classical formulas); this supposition, although plausible, has not been tested by experiment.

There are two situations in which the activation-in-many-degrees-of-freedom rate constant would be useful in bringing theory and experiment into agreement.   The first would be a situation in which using the line-of-centers rate constant would require $p > 1$.   The second would be a case in which the experimental activation energy is less than the minimum activation energy deduced from the thermochemistry of the reaction.   In either case, a value of s is found, using the methods described in Section 3-4, which brings agreement between the collision theory and the experimental measurement.

EXERCISE 4-7   The elementary gas reaction $H + H_2O \rightarrow OH + H_2$ is endothermic by 15 kcal.   Calculate the upper limit value for $k(2,000°K)$ from (a) the line-of-centers $k(T)$ formula, and (b) the activation-in-many-degrees-of-freedom $k(T)$ formula.   Use $\sigma = 2.0$ Å and assume the value of s to be 6.

EXERCISE 4-8   The reactive cross section for the ion–molecule reaction $N_2^+ + N_2 \rightarrow N_3^+ + N$ was found to have a threshold energy of 9 eV and a maximum cross section of 10^{-17} cm^2.   Assuming that $Q(E, T)$ for $E > E_0 = 9$ eV varies as $Q(E, T) = \alpha(E - E_0)^2 \exp[-\beta(E - E_0)]$ and that the maximum cross section is at $E = 30$ eV, find the corresponding $k(T)$ function (cf. Exercise 4-6).

EXERCISE 4-9   Show that the cross section $Q(E, T) = 0$, $E < E_0$, and $Q(E, T) = c(E - E_0)^{\frac{1}{2}} E^{-1}$, $E > E_0$ leads to a rate constant expression of the Arrhenius form.

EXERCISE 4-10 Reactions between ions and molecules are often described by cross sections that stem from the electrostatic forces attracting the reactants to one another. The ion-induced dipole cross section (Langevin cross section) as a function of relative velocity is $Q(u) = 2\pi e(\alpha/\mu)^{\frac{1}{2}} u^{-1}$. Show that the corresponding rate constant is $k = 2\pi e(\alpha/\mu)^{\frac{1}{2}}$. You may find it helpful to use an integral over u rather than an integral over E. In these equations e is the charge on the ion, α is the polariz-·ability of the molecule, and μ is the reduced mass of the pair.

4-4 POTENTIAL ENERGY SURFACES

So far we have accepted the idea that A + BC encounters occur and can be characterized by a reactive cross section without inquiring into the motions of atoms A, B, and C during the encounter. Now we consider a dynamical theory that can be used to investigate these motions.

While A and BC are still far apart from one another their translational motion is straight ahead; that is, the velocity vectors $\mathbf{u_A}$ and $\mathbf{u_{BC}}$ are constant in time. The internal motion of BC is a combination of the vibrational motion determined by its v state and the rotational motion determined by its J state. As A and BC come close to one another, however, they interact with one another and the atomic motions change. Interatomic forces cause the velocity vectors $\mathbf{u_A}$ and $\mathbf{u_{BC}}$ to change, and the motions of B and C are no longer those of an isolated BC molecule. If we set out to describe these changes in the motions of A, B, and C by Newtonian mechanics, we would say that A, B, and C are subject to accelerations by the law of motion $\mathbf{f} = m\mathbf{a}$, where $\mathbf{f}$ represents the net force that one of the atoms experiences in the vicinity of the other two. It is more convenient, however, to describe the interactions among the atoms in terms of a *potential function V* from which the forces on the individual atoms can be derived according to the usual rule of mechanics

$$f_{Ax} = -\frac{\partial V}{\partial x_A}$$

where f_{Ax} is the x component of the force experienced by atom A.

In order to find the x, y, and z components of the forces on A, B, and C it might seem at first that we would need nine such partial derivatives, and V would, therefore, have to be a function of the *nine* variables x_A, y_A, z_A, x_B, y_B, z_B, x_C, y_C, and z_C. This is more information than is really desired, for these nine coordinates locate the c.m. of the ABC system, which could be anywhere in the reaction vessel as far as we are concerned. All that is pertinent to the *relative motion* of A, B, and C is the dependence of the potential function V on the *relative positions* of A, B, and C and on the motion of the plane containing A, B, and C. One way to express the relative positions would be in terms of a pair of vectors, for example the vector $\mathbf{r}_{AB}$ connecting A to B and the vector $\mathbf{r}_{BC}$ connecting B to C; the relative position vector $\mathbf{r}_{AC}$ connecting A and C would be equal to the sum of the first two, or $\mathbf{r}_{AC} = \mathbf{r}_{AB} + \mathbf{r}_{BC}$. The relative positions are thus a function of *six* variables, namely, the components of $\mathbf{r}_{AB}$ and $\mathbf{r}_{BC}$. The six variables need not be these components, however. Suppose that we choose three of the six to be the scalar distances between the atoms, that is, r_{AB}, r_{BC}, and r_{AC}. Then the three remaining variables can be chosen to describe the motion of the plane containing A, B, and C. If this plane is stationary for a particular A + BC encounter, then the only potential energy involved will be that depending on the interatomic distances, and $V = V(r_{AB}, r_{BC}, r_{AC})$. If this plane is not stationary for a particular A + BC encounter, then the atoms are subject to centrifugal as well as interatomic forces during the encounter. If we denote the orientation variables as α, β and γ, the effective potential energy during the encounter will depend on the rate of change of orientation of the ABC plane as well as on the interatomic distances, and we would have an effective potential energy function $V = V(r_{AB}, r_{BC}, r_{AC}, \dot{\alpha}, \dot{\beta}, \dot{\gamma})$. This separation of the effective potential energy function into contributions from interatomic forces and centrifugal forces is the three-particle equivalent of using different potential energy curves for different rotational states of a diatomic molecule (Figure 5-2).

Considering centrifugal contributions to the potential energy surface is necessary for accurate calculations, but not necessary for understanding the use of potential energy surfaces in chemical kinetics. Let us therefore be content to study the case where the ABC plane is stationary, so that we may write $V = V(r_{AB}, r_{BC}, r_{AC})$. Suppose that the reaction of interest is endothermic ($\Delta H > 0$) and has activation energy $E_A > \Delta H$. When A is far away from BC, r_{AB} and r_{AC} will be large, while r_{BC} oscillates about the equilibrium internuclear distance of BC. The forces between A and BC are negligible. Let us arbitrarily call this potential energy $V = 0$. As A approaches BC, it encounters *repulsive* forces (hence the need for relative kinetic energy to overcome E_A) and V *increases*. This increase in V is a barrier to reaction, and the minimum increase in V in an A + BC encounter that could lead to reaction is called the *barrier height* E_b. It is approximately equal to the activation energy E_A and the threshold energy E_0. After reaction occurs, the products AB and C separate; r_{AB} oscillates about the equilibrium internuclear distance of AB while r_{BC} and r_{AC} become large. During the separation AB and C repel one another; V decreases and eventually becomes approximately equal to ΔH. The exact energy relationships taking zero-point vibrational energies into account are shown in Figure 4-4.

It is not possible to draw a graph of $V(r_{AB}, r_{BC}, r_{AC})$ because this would require a four-dimensional space. For some A + BC reactions, however, semiempirical quantum mechanical calculations suggest that the most favorable encounter—the one with the smallest E_b—will be a *collinear* one, with A, B, and C lying on a straight line. If we limit ourselves to this special case, then V becomes a function of *two* independent variables only, since any r_{ij} can be found from the other two, and the potential energy function is indeed a surface in the usual sense of the word, as shown in Figure 4-4. Let us choose r_{AB} and r_{BC} as the independent variables. Large values of r_{AB} correspond to the separated reactants A and BC; large values of r_{BC} correspond to the separated products AB and C. The main qualitative features of a $V(r_{AB}, r_{BC})$ surface are the same as those of a $V(r_{AB}, r_{BC}, r_{AC})$ surface. In Figure 4-4 a $V(r_{AB}, r_{BC})$ surface calculated for the reaction $O + H_2 \rightarrow OH + H$ by empirical methods is shown.

The potential energy surface of Figure 4-4 represents still another way of thinking about bimolecular gas reactions. From a quantum mechanical

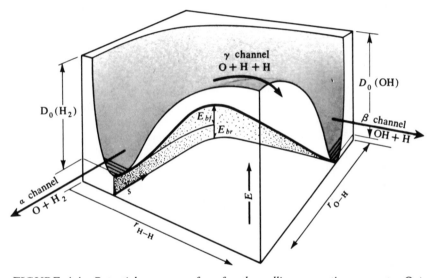

FIGURE 4-4 *Potential energy surface for the collinear reactive encounter* $O + H_2 = OH + H$. *Experimental data for the* H_2 *and* OH *molecular potential curves were obtained in accurate spectroscopic experiments. The activation energy for this reaction is about* 10 kcal, *from which a barrier height for the forward reaction* E_{bf} *of about* 10 kcal = 3,500 cm^{-1} *can be estimated. The heat of reaction* ΔH_0 *is* -1.9 kcal $= -670$ cm^{-1}. *Reaction into the* γ *channel to give three atoms would not be observed experimentally owing to the high energy necessary; that is, for* $O + H_2$ *energies less than* $D_0(H_2)$, *the* γ *channel is closed. Reactive encounters originating from the* γ *channel are, in principle, observable, but would be complicated by a fourth channel (not shown here) corresponding to an excited electronic state of* OH *with energy* 32,720 cm^{-1} *greater than the ground electronic state. Reactions that occur on a single potential surface are called adiabatic; reactions described by crossings between surfaces are called nonadiabatic reactions.*

point of view the cross section for reactive scattering in a collinear A + BC encounter would be calculated by finding solutions of the time-dependent Schrödinger equation that describes the motion of a wave packet representing the A—B—C system on the $V(r_{AB}, r_{BC})$ potential energy surface. This is the basis of the quantum theory of scattering alluded to at the beginning of this chapter. Unfortunately, this Schrödinger equation has only been solved by approximate methods, and the quantum theory of scattering has not yet contributed strongly to chemistry. An alternative way of gaining provisional information about an A + BC encounter is to assume that the important features of the encounter are predicted just as well by classical mechanics as by quantum mechanics. If this assumption is made, then the course of any A + BC encounter can be calculated exactly; one substitutes an exact solution of the possibly nearly valid classical mechanical problem for an approximate solution of the completely valid quantum mechanical problem. The justification for doing this is essentially that the potential changes only slightly over a distance corresponding to the deBroglie wavelengths of translational motion in a typical A + BC encounter.

To obtain a classical mechanical description of a single A + BC encounter, the differential equations of motion are integrated numerically using a digital computer. The computer output tells the positions of A, B, and C as time progresses. This information, called the *trajectory* of the encounter, tells whether reaction occurs and also describes how reaction occurs. In a trajectory study for a collinear encounter a large number of numerical integrations are carried out for a range of initial conditions of relative velocity and vibrational phase using a mathematical procedure known as the Monte Carlo method to select a random set of initial conditions. By taking appropriate averages and weighting with Boltzmann factors as required, a reactive cross section function $Q(E, T)$ can be determined. Since the computer is just as content—albeit more demanding in time—with $V(r_{AB}, r_{BC}, r_{AC}, \dot{\alpha}, \dot{\beta}, \dot{\gamma})$ as it is with $V(r_{AB}, r_{BC})$, a more ambitious trajectory study can be done for the full range of A-B-C orientations and without constraining the ABC plane to be stationary. Averaging over the rotational and vibrational energies, relative energy, and impact parameter again provides a $Q(E, T)$ function.

Trajectory studies provide several kinds of information besides reactive cross sections $Q(E, T)$. For example, by varying the potential energy surface one can discover which characteristics of the surface are important for determining whether energy released in a reaction will appear in relative translational energy of AB and C or in vibrational energy of AB. Also, one can discover whether the trajectories for a given surface are *simple*—A encounters BC and leaves with B before A, B, and C together carry out a large number of rotations and oscillations—or *tangled*—A, B, and C do stay together long enough to carry out rotations and oscillations before AB and C part. These dynamical results from trajectory studies are relevant to certain experiments (Section 4-5) and to theories of bimolecular (Section 4-7) as well as unimolecular and termolecular (Chapter 5) gas reactions.

4-5 MEASUREMENT OF REACTIVE CROSS SECTIONS

We found in Section 4-2 that explicit equations relating $k(T)$ to $Q(E, T)$ could be derived if certain functional forms for $Q(E, T)$ were assumed. Accepting these functional forms therefore permits experimental rate constants for bimolecular gas reactions to be converted to the corresponding reactive cross sections. As mentioned before, however, there are good reasons for believing that no single functional form describes $Q(E, T)$ for all bimolecular gas reactions; there are also good reasons for believing that no simple $Q(E, T)$ functions accurately describe the reactive cross section for any bimolecular gas reaction. In these circumstances, it is appropriate to ask whether experimental rate constants provide any information at all about the functional forms of reactive cross sections. The answer to this appears to be no. This is due to the way in which $Q(E, T)$ combines with the distribution function $E \exp(-E/kT)$ in the integrand of the fundamental $k(T)$ equation. In chemical kinetics experiments, reaction rates are usually measured at temperatures where RT is much less than E_A and therefore much less than E_0. For such temperatures the factor $E \exp(-E/kT)$ is a rapidly decreasing function for E near E_0. As shown in Figure 4-3, the integral of the fundamental $k(T)$ equation is therefore a product of one function that decreases rapidly near E_0 and another function, $Q(E,T)$, that increases rapidly near E_0. This means that the value of the entire integral is determined by the value of the product $E \exp(-E/kT)Q(E, T)$ in the immediate vicinity of the threshold energy E_0. Since the rate constant is determined by the behavior of the cross section function $Q(E, T)$ over such a narrow range of E, there is no way of utilizing information about $k(T)$ to obtain information about $Q(E, T)$ for E beyond the threshold energy E_0. All that $k(T)$ data tell about $Q(E,T)$ is the approximate value of E_0—from E_A—and the value of the integral over all E.

In order to gain information on $Q(E, T)$ functions, or $Q(E)$ functions for specific quantum states, we must, therefore, consider experiments in which the reactants are not in a Boltzmann distribution, so that the rate of product formation is not governed by an equation containing $E \exp(-E/kT)$. Several experimental methods achieve this. We mention two of them briefly here, and discuss these two and others at greater length in Chapters 9 and 10. The first requirement is to achieve $E > E_0$ by variable, known amounts; the second is to measure the reactive cross section as a function of E.

One way of obtaining high relative energy encounters is to heat one of the reactants to an extremely high temperature in a furnace. This reactant is allowed to diffuse out of the furnace and is collimated into a beam of molecules with high translational velocity. The second reactant is introduced as a second beam intersecting the first at a right angle. Detectors sensitive to product molecules measure $Q(E, T)$, where E is determined mainly by the high velocity of molecules or atoms in the beam coming from the furnace. This is the *molecular beam* experiment, further described in Section 9-6.

A second way of obtaining high relative energy encounters is to irradiate appropriate molecules with uv light, thus causing bond ruptures from which the fragment atoms or molecules recoil at high velocity. If these *hot atoms* (or molecules) react in their first few encounters, they can be used to measure $Q(E, T)$ for $E \gg E_0$. Quantitative analysis of the reaction products, corrected by mathematical analysis to take account of elastic collisions that can precede reaction, measures the values of $Q(E, T)$, and varying the wavelength of the irradiating light serves to vary E. Further description of the hot-atom experiment is given in Section 10-5.

4-6 ENERGY TRANSFER

We have not paid any attention so far to bimolecular encounters in which reaction does not occur. The study of cross sections for the inelastic channels corresponding to changes in electronic, rotational, or vibrational quantum states is also relevant to chemical kinetics, however, and comprises the gas phase portion of the subject of *energy transfer*. A wide variety of experimental techniques provide energy transfer cross-section information. Among them are shock tube experiments (Section 9-5), flash photolysis (Section 9-4), ultrasound (Section 9-2), and fluorescence quenching (Section 10-2). Here we summarize the results that are important for chemical kinetics in gas reactions.

Energy transfer channels are named according to type of energy transferred, with the letters E, R, V, and T standing for electronic, rotational, vibrational, and translational, respectively. Transfer of energy from vibration to translation is VT transfer, from vibration to rotation is VR transfer, and so on. When internal states (E, V, R) transfer energy to other internal states, as in VR transfer, the energy differences caused because the quanta are incommensurate are necessarily made up by translational energy.

Rotational energy transfer, both RT and RR, is found to occur with large cross sections, typically 1–100 $Å^2$. This implies that Boltzmann distributions of rotational states are maintained effectively in gases.

Vibrational energy transfer is found to occur with large (10–100 $Å^2$) cross sections for encounters between large molecules (VV transfer), but with small (10^{-5}–10^{-1} $Å^2$) VT cross sections for small molecules, in particular diatomic molecules. For reactions involving small molecules, therefore, there are experimental conditions in which a Boltzmann distribution in vibrational states does not apply. The efficiency of VV transfer in encounters between large molecules is a cornerstone of the theory of unimolecular reactions developed in the next chapter.

Cross sections for transfer of electronic energy vary greatly, depending on the specific atoms or molecules involved. Some examples are given in Section 10-2.

4-7 ACTIVATED COMPLEX THEORY

We consider now a theory that, through adoption of some special postulates, permits prediction of $k(T)$ functions from molecular properties. Although this theory also encompasses unimolecular and termolecular gas reactions, and solution reactions as well, it is most clearly formulated for bimolecular gas reactions. It focuses attention on the pair of reactants at the saddle point of the potential energy surface; because this highest energy configuration is at the saddle point of a rather complicated multidimensional potential energy surface it is called an *activated complex*, from which the activated complex theory (ACT) derives its name. In ACT it is postulated that activated complexes are in equilibrium with the reactants, and that the concentration of activated complexes can be calculated from an equilibrium constant.

Specializing our argument temporarily to consider the collinear A + BC surface of Figure 4-4, the curved line connecting the lowest points of the A + BC and AB + C valleys over the saddle point represents the least energetic pathway for the reaction (ignoring for the moment the complication due to zero-point vibrational energies of AB and BC). In ACT it is assumed that a unique minimum energy pathway such as this exists for all bimolecular reactions. The minimum energy pathway is called the *reaction coordinate*. (For cases other than the collinear A + BC reaction it is still assumed to be a line, but in a multidimensional space.) It is always identified with a specific molecular motion (the H-atom transfer in the case depicted in Figure 4-4), which is postulated in ACT to be separable both conceptually and computationally from all other motions, translational as well as internal, of the two reactant molecules.

Let us generalize our schematic bimolecular reaction to

A + B → products

where A and B are any reactant atoms or molecules, and let us denote the postulated activated complex by C. We express the postulated equilibrium between reactants and activated complex by writing the mechanism

A + B ⇌ C → products

with the equilibrium constant

$$K_N^\dagger \equiv \frac{{}^1 N_C}{{}^1 N_A \, {}^1 N_B}$$

where concentrations are given in molecules cm^{-3}. This equilibrium constant is expressed in terms of partition functions per unit volume q_A^*, q_B^*, and q_C^*, each referred to their individual zero of energy, and a Boltzmann factor $\exp(-E_b'/kT)$, which adjusts q_A^*, q_B^*, and q_C^* to a common energy zero

$$K_N^\dagger = \frac{q_C^*}{q_A^* q_C^*} e^{-E_b'/kT}$$

An energy level diagram illustrating the correction to a common energy zero is shown in Figure 4-5.

[This statistical mechanical formula can be made plausible by the following simple argument. In a 1-cm³ sample containing gases A, B, and C the relative probability of finding an activated complex C rather than a pair A + B is

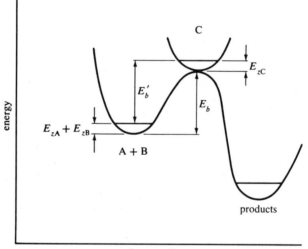

FIGURE 4-5 *Schematic diagram indicating the accounting procedure for zero-point energy in activated complex theory calculations. The barrier height E_b is the same quantity shown in Figure 3-1. The energy E_b' is the barrier height corrected for the zero-point energies of the reactants and the activated complex by the formula $E_b' = E_b + E_{zC} - E_{zA} - E_{zB}$. Each of the zero-point energies E_z is given by a sum over the appropriate set of vibrational frequencies $E_z = \frac{1}{2} \sum h\nu_i$.*

given by the ratio of the Boltzmann factors

$$\exp(-E_{C,i}/kT) \text{ and } \exp(-E_{A,B,j}/kT)$$

summed over all possible quantum states i and j of reactants A and B and complex C. These sums are the q^*'s of this equation after $\exp(-E'_b/kT)$ is factored out.]

One of the degrees of freedom of C is the reaction coordinate. If C were a normal molecule, this would be a vibrational degree of freedom. In an activated complex, however, this is considered to be a translation. By postulate, motion in the reaction coordinate is separable from the other degrees of freedom of C. The partition function q_C^* can therefore be factored into the translational partition function in the reaction coordinate $(2\pi\mu kT)^{\frac{1}{2}}\delta/h$, where μ is the reduced mass along the reaction coordinate and δ is a small unit of length along the reaction coordinate, and the remaining partition function $q_C^{*\dagger}$. This gives for ^1N_C

$$^1N_C = \frac{(2\pi\mu kT)^{\frac{1}{2}}}{h} \delta \frac{q_C^{*\dagger}}{q_A^* q_B^*} e^{-E'_b/kT} \, ^1N_A \, ^1N_B$$

in molecules cm^{-3}. The number of activated complexes per unit length of reaction coordinate, called the density of activated complexes, is given by the derivative of ^1N_C with respect to δ, or

$$^1N'_C = \frac{(2\pi\mu kT)^{\frac{1}{2}}}{h} \frac{q_C^{*\dagger}}{q_A^* q_B^*} e^{-E'_b/kT} \, ^1N_A \, ^1N_B$$

in molecules cm^{-4}. Half of these activated complexes are moving in one direction on the reaction coordinate while half are moving in the other. The rate of motion along the reaction coordinate is assumed to be given by the kinetic theory of gases formula for the average velocity in one direction of a particle of mass μ

$$\langle v \rangle = (2kT/\pi\mu)^{\frac{1}{2}}$$

Dividing the $^1N'_C$ formula by 2 to allow for the two possible directions of motion and multiplying it by $\langle v \rangle$ gives the average rate at which activated complexes become products

$$R = (2\pi\mu kT)^{\frac{1}{2}} \frac{(2kT/\pi\mu)^{\frac{1}{2}}}{2h} \frac{q_C^{*\dagger}}{q_A^* q_B^*} e^{-E'_b/kT} \, ^1N_A \, ^1N_B$$

or, in cm^3 molecule^{-1} sec^{-1}

$$k(T) = \frac{kT}{h} \frac{q_C^{*\dagger}}{q_A^* q_B^*} e^{-E_b'/kT}$$

This is the ACT equation for $k(T)$. It has the apparent advantage over, for instance, the line-of-centers cross-section formula, that all of the quantities on the right-hand side of the equation are derivable in principle from known properties of stable molecules.

EXERCISE 4-11 Compare the line-of-centers formula for $k(T)$ with the ACT formula to show that for $E_0 = E_b'$ the reactive cross section $Q(E,T) = p\pi\sigma^2$ is given in ACT by

$$Q(E, T) = \left(\frac{\pi\mu kT}{8h^2}\right)^{\frac{1}{2}} \frac{q_C^{*\dagger}}{q_A^* q_B^*}$$

The partition functions per unit volume can be calculated for ACT using the ideal gas, rigid rotor, harmonic oscillator approximation to the molecular energy levels. (For the reactants A and B it may be a simpler matter to calculate q^* using handbook tables of the free energy function, $(\bar{G}_T^0 - \bar{H}_0^0)/RT$, and the statistical mechanics formula $q^* = 7.34 \times 10^{21} \; T^{-1} \; \exp[-(\bar{G}_T^0 - \bar{H}_0^0)/RT]$ cm^{-3}.) Calculating $q_C^{*\dagger}$, however, always involves devising a model of the activated complex and doing a direct calculation of $q_C^{*\dagger}$. This calculation is one crux of obtaining $k(T)$ in ACT, the other obviously being the calculation of E_b'. Choosing a model activated complex means estimating its shape, its size, and the frequencies of its molecular vibrations. The shape and size determine the moment(s) of inertia and hence the rotational energy levels. The set of molecular vibration frequencies determines the vibrational energy levels.

The calculation of $q_C^{*\dagger}$ proceeds by factoring into a product of translational, electronic, rotational, and vibrational partition functions

$$q_C^{*\dagger} = q_{tC}^* q_{eC} q_{rC} q_{vC}$$

The first three partition functions give no special computational problems. We collect here the statistical mechanical formulas that are used. The translational partition function depends on the mass of the activated complex m_C and on the temperature

$$q_{tC}^* = \left(\frac{2\pi m_C kT}{h^2}\right)^{\frac{3}{2}}$$

The electronic partition function is a constant equal to the electronic degeneracy of the activated complex

$$q_{eC} = g_C$$

The rotational partition function depends on the moment of inertia I_C for a linear activated complex and on the product of inertia $I_x I_y I_z$ for a nonlinear activated complex

$$q_{rC} = \frac{8\pi^2 I_C kT}{\sigma h^2} \qquad \text{linear}$$

and

$$q_{rC} = \frac{\pi^{\frac{1}{2}}}{\sigma} \left(\frac{8\pi^2 (I_x I_y I_z)^{\frac{1}{3}} kT}{h^2} \right)^{\frac{3}{2}} \qquad \text{nonlinear}$$

The moments of inertia as well as the symmetry number σ are obtained from the assumed geometry of the activated complex. This geometry is derived by assuming that bond angles and bond lengths in activated complexes are similar to bond lengths and bond angles in similar stable molecules. (*Warning: For some activated complexes the symmetry number must be calculated by special rules in order to account for multiple reaction paths.*)

Calculation of q_{vC} poses a serious problem.

A linear N-atomic molecule has 3 translational, 2 rotational, and $3N - 5$ vibrational degrees of freedom. A nonlinear N-atomic molecule has 3 translational, 3 rotational, and $3N - 6$ vibrational degrees of freedom. If the molecule happens to be an activated complex, one of the vibrational degrees of freedom is the reaction coordinate, however, and there are therefore only $3N - 6$ vibrational degrees of freedom for a linear activated complex and $3N - 7$ vibrational degrees of freedom for a nonlinear activated complex. In the harmonic oscillator approximation, the vibrational partition function is a product over $3N - 6$ (or $3N - 7$) factors $(1 - \exp[-\Theta_{v,i}/T])^{-1}$, where the characteristic temperature for the ith vibrational degree of freedom with fundamental frequency v_{0i} sec^{-1} is $\Theta_{v,i} = h v_{0i}/k$. Thus

$$q_{vC} = \prod_{i=1}^{3N-6 \text{ or } 7} (1 - \exp[-\Theta_{v,i}/T])^{-1}$$

The set of frequencies $\{v_{0i}\}$ for the activated complex is not known and must be estimated by analogy with vibrational frequencies for stable molecules. For stiff, high frequency vibrations this is no problem; since stiff vibrations have large $\Theta_{v,i}$, and the corresponding factors in q_{vC} are near unity, only small errors in q_{vC} are made if v_{0i} values for stiff vibrations are incorrectly estimated. Unfortunately, the activated complex necessarily has weak vibra-tions as well, in particular the bending vibration perpendicular to the reaction coordinate. Estimates of these frequencies are usually the main source of probable error in calculations of $q_C^{*\dagger}$.

The value of E_b' to be used includes the difference in zero-point energies for A, B, and C, since the vibrational partition functions use the $v = 0$ level as the zero of energy rather than the potential minima, as indicated in Figure 4-5. For an ordinary equilibrium constant calculation in statistical mechanics this is simply the heat of reaction at zero degrees Kelvin. For a calculation of $K_N^\dagger$, however, this heat of reaction is not known. Empirical rules relating E_b' to the pertinent bond strengths are used instead.

Deriving expressions for $k(T)$ in the ACT formalism thus requires two uncertain procedures: estimating the set of vibrational frequencies of the activated complex and estimating the barrier height. Success in using ACT to predict $k(T)$ depends on choosing the right procedures for making these estimates.

The theoretical foundations of ACT are actually somewhat more secure than they would appear to be from the above presentation. The ACT formula for $k(T)$ can be derived from several more general theoretical ap-proaches when simplifying assumptions are made. It appears that the really awkward fundamental difficulty with the foundations of ACT is the doubtful basic premise that an identifiable and computationally separable reaction coordinate exists for each gas phase bimolecular reaction. In the special case of a collinear A—B—C activated complex, the separability assumption is obviously valid. The reaction coordinate here would be the "asymmetric stretch" normal mode of vibration for a stable linear ABC molecule such as CO_2. It may be quite reasonable to identify some reaction coordinates with the asymmetric stretch of an unstable linear activated complex and to treat them computationally as separable degrees of freedom. It is difficult to believe, however, that such facile identification and computational separability are really valid for elementary reactions whose transition states are not linear and triatomic. This "separability problem" has led theoreticians to seek alternative quantum mechanical or quantum statistical procedures for calculating bimolecular reaction rate constants. Unfortunately, these alternatives have not yet met with general success.

In addition to the separability problem, the serious difficulties encountered in calculating E_b' and q_{vC} force one to regard all ACT rate constant calculations with some degree of skepticism.

In Chapter 5 we use a procedure similar to ACT as part of the theory of unimolecular reactions. In Chapter 6 we develop a thermodynamic for-mulation of ACT that will prove to be more suitable than the statistical mech-anical formulation for interpretation of solution reactions.

EXERCISE 4-12 Select one of the atom + diatomic molecule reactions from Table 4-1 and devise a model for the activated complex. Write out the ACT rate constant formula for your model. Explain how you would obtain values for all the parameters appearing in your formula.

In deriving the ACT rate constant formula we calculated the rate of decomposition of activated complexes on the assumption that there was a separable reaction coordinate on which translational motion occurred. Let us now suppose that the ACT formalism correctly predicts the concentration of activated complexes, but that motion along the reaction coordinate is somehow hampered. The activated complexes do not decompose as rapidly as we have calculated before on the basis of free translational motion. We multiply the ACT rate constant formula by some function $f(T)$ that corrects it for the failure of the free translation assumption, whereupon the rate constant formula becomes

$$k(T) = f(T) \, \frac{kT}{h} \, \frac{q_C^{*\dagger}}{q_A^* q_B^*} \, e^{-E_b'/kT}$$

Whatever the function $f(T)$ might be, it will be necessary to multiply the rate constant for the reverse reaction by the same function in order that the ratio of the formulas for the forward and reverse rate constants will still give the correct equilibrium constant expression.

Admitting the possibility that a correction function of this kind could be appropriate introduces an additional degree of arbitrariness into the ACT formalism. It was realized early in the development of ACT that this arbitrariness might be unavoidable. The proposed correction function was a simple constant κ, called the *transmission coefficient*, having a value in the range $0 < \kappa < 1$. This would give a revised ACT rate constant formula

$$k(T) = \kappa \, \frac{kT}{h} \, \frac{q_C^{*\dagger}}{q_A^* q_B^*} \, e^{-E_b'/kT}$$

Unfortunately, there is no way to do an *a priori* theoretical calculation of the value of κ for any reaction. For this reason the transmission coefficient is often regarded as a loophole for bringing ACT calculations into agreement with experimental rate constants that turn out to be smaller than indicated by the ACT calculations. With the advent of large-scale trajectory calculations (Section 4-4) it became possible to find the rate of decomposition of activated complexes on potential energy surfaces for which ACT calculations could also be made. For the reaction $H + H_2 \rightarrow H_2 + H$, the value of κ found by comparing the trajectory calculations to the ACT calculations is close to unity. The situation for other reactions remains to be investigated.

EXERCISES

4-13 Write out a derivation showing that $Q(E)$ could be defined through the equation $k(E) = uQ(E)$, where u is the relative velocity $u = (2E/\mu)^{\frac{1}{2}}$, $k(E)$ is a rate constant for relative energy E, and internal quantum states are ignored. Your derivation should show that the fundamental equation relating the macroscopic rate constant $k(T)$ to $Q(E)$ has the same form for this definition of $Q(E)$ as it does for $Q(E)$ defined through the reactive scattering argument used in this chapter. Rate constants that are functions of molecular-scale variables, as $k(E)$ here, play prominent roles in the theories of chemical kinetics and usually called *microscopic* or *specific* rate constants.

4-14 Explain the relationship between the potential energy surface of Figure 4-4 and the energy diagram of Figure 3-1.

4-15 Show that the average velocity in one direction for a particle of mass μ is $\langle v \rangle = (2kT/\pi\mu)^{\frac{1}{2}}$ by evaluating

$$\int_{-\infty}^{\infty} |u_{Ax}| f(u_{Ax}, u_{Ay}, u_{Az})\, du_{Ax}\, du_{Ay}\, du_{Az}$$

$$= 2 \int_{0}^{\infty} u_{Ax} f(u_{Ax}, u_{Ay}, u_{Az})\, du_{Ax}\, du_{Ay}\, du_{Az}$$

The distribution function can be found in Section 4-1.

4-16 A linear approximation is sometimes used for describing atom–atom collisions near threshold. (a) Show that $Q(E) = a(E - E_0)$ leads to the rate constant formula

$$k(T) = a(8(kT)^3/\pi\mu)^{\frac{1}{2}}(E_0/kT + 2)\exp(-E_0/kT)$$

(b) The cross section for exciting an H atom to its first electronically excited state (principal quantum number = 2) by a collision with another H atom was found in shock tube experiments (Section 9-5) to be $Q(E) = 1.3 \times 10^{-18}(E - 10.2)$ cm^2, where E is the relative energy in electron volts and 10.2 eV is the threshold energy for excitation. Calculate the rate constant for this process at $T = 12{,}000°$K.

4-17 The average time required for a free electron to recombine with an ion in H II regions near the hottest stars of spiral arms of galaxies, where the ion and electron densities are about 10^3 cm^{-3}, is about 10 years. (a) Estimate the second order rate constant in cm^3 sec^{-1} units for the principal reaction responsible for deionization

$$e^{(-)} + p^{(+)} \rightarrow H^\dagger + h\nu$$

where $e^{(-)} =$ electron, $p^{(+)} =$ H II $=$ proton, and $H^\dagger =$ an electronically excited hydrogen atom. (b) Calculate the reactive cross section, assuming $E_0 = 0$ and $T = 5{,}000°$K.

SUPPLEMENTARY REFERENCES

An extensive collection of experimental data on bimolecular gas reactions can be found in

A. F. Trotman-Dickenson and G. S. Milne, *Tables of Bimolecular Gas Reactions*, National Bureau of Standards Rept. NSRDS–NBS 9, 1967.

A critical survey of rate constant data for selected bimolecular gas reactions is given by

K. Schofield, *Planetary and Space Science* **15**, 643 (1967).

Bimolecular gas reactions between ions and molecules are discussed in a recent symposium volume

P. J. Ausloos, Ed., *Ion–Molecule Reactions in the Gas Phase; Advances in Chemistry Series 58* (American Chemical Society, Washington, D.C., 1966).

An introductory description of collision theory from a viewpoint similar to that taken in this chapter is

E. F. Greene and A. Kupperman, "Chemical Reaction Cross Sections and Rate Constants," *J. Chem. Ed.* **45**, 361 (1968).

The activated complex theory is described by

S. Glasstone, K. J. Laidler, and H. Eyring, *Theory of Rate Processes* (McGraw-Hill, New York, 1941).

Energy transfer theory and experiments are discussed by

A. B. Callear, "Energy Transfer in Molecular Collisions," in *Photochemistry and Reaction Kinetics*, P. G. Ashmore, F. S. Dainton, and T. M. Sugden, Eds. (Cambridge University Press, Cambridge, England 1967).
T. L. Cottrell, *Dynamic Aspects of Molecular Energy States* (Oliver and Boyd, London, 1965).
W. H. Flygare, "Molecular Relaxation," in *Accounts of Chem. Res.* **1**, 121 (1968).
R. G. Gordon, W. Klemperer, and J. I. Steinfeld, "Vibrational and Rotational Relaxation," in *Annual Reviews of Physical Chemistry Volume* 19, H. Eyring, C. J. Christensen, and H. S. Johnston, Eds., (Annual Reviews Inc., Palo Alto, Calif., 1968).
B. Stevens, *Collisional Activation in Gases* (Pergamon, London, 1968).

A classical exposition of the collision theory, including derivations of the necessary statistical mechanical formulas, is given by

L. S. Kassel, *The Kinetics of Homogeneous Gas Reactions* (Chemical Catalog Company, New York, 1932), Chapters II and III.

Tables of collision diameters can be found in

J. O. Hirschfelder, C. F. Curtiss, and R. B. Bird, *Molecular Theory of Gases and Liquids* (J. Wiley & Sons, New York, 1954).

To obtain hard sphere collision diameters σ_{HS} from the Lennard–Jones parameters σ listed in Table I-A of this reference, use the following procedure:
(1) Calculate the average inverse temperature for the temperature range of the experiments being considered from the formula

$$\overline{T} = T_1 T_2 / (T_1 + T_2)$$

(2) Find ε/k for each of the molecules in Table I-A;
(3) Form $T^* = \overline{T}k/\varepsilon$ for each molecule;
(4) Find $\Omega^{(2,2)\star}$ for each molecule by interpolation in Table I-M;
(5) Calculate σ_{HS} for each molecule from the values of σ in Table A-I and

$$\sigma_{HS} = \sigma^2 \Omega^{(2,2)\star}$$

(6) Calculate σ_{HS} for the pair of molecules from

$$\sigma_{HS} = \tfrac{1}{2}[\sigma_{HS}(\text{molecule 1}) + \sigma_{HS}(\text{molecule 2})]$$

Potential energy surfaces, trajectory calculations, and activated complex theory are treated by

D. L. Bunker, *Theory of Elementary Gas Reaction Rates* (Pergamon Press, Oxford, 1966), Chapters 1 and 2.

H. S. Johnston, *Gas Phase Reaction Rate Theory* (Ronald Press Co., New York, 1966).

K. J. Laidler and J. C. Polanyi, "Theories of the Kinetics of Bimolecular Reactions," in *Progress in Reaction Kinetics—Volume 3* (Pergamon Press, Oxford, 1965), pp. 1–61.

Chapter 5

UNIMOLECULAR AND TERMOLECULAR

GAS REACTIONS

ONE OF the first concepts introduced in a college physics course is mechanical energy, comprised of kinetic energy, the $mv^2/2$ term attributed to motion, and potential energy, caused by such things as forces arising from gravitational fields or from springs, or by interactions of charged objects or magnetized objects with fields or with other charged objects or magnetized objects. The concept of mechanical energy is just as important in theories of chemistry as it is in theories of physics. In molecular quantum mechanics one deals with the energy of bonding electrons and considers the stability of chemical bonds and the spectroscopically observed transitions between energy levels. In statistical thermodynamics one deals with the distribution functions of molecules in energy levels and computes from these the values of molar thermodynamic properties. In chemical kinetics, and particularly in the theory of gas reactions, we are interested in the subject of *energy transfer* and in the dependence of reactivity on energy. For bimolecular reactions, energy transfer refers to interchange of translational and vibrational or rotational energy that results from encounters, for example, of an atom with a diatomic molecule, or to the distribution functions in translational and internal energy of the products of a reactive encounter (cf. Section 4-6). Energy transfer plays an even more important role in the theory of unimolecular and termolecular gas reactions. In this chapter we consider the gas kinetics experiments that provide information about intermolecular energy transfer in bimolecular or termolecular encounters and about intramolecular migration of vibrational energy within single vibrationally "energized" molecules. The dependence of reactivity on degree of energization also plays an important role in the theory of unimolecular reactions, analogous to the dependence of bimolecular reactive cross sections on relative kinetic energy.

5-1 MECHANISM OF FIRST ORDER GAS REACTIONS

In the early history of gas kinetics it was uncertain whether first order gas reactions existed and unclear how one should interpret the rate constant of a unimolecular elementary reaction. To obtain unequivocal evidence to settle the experimental question, researchers concentrated their attention on thermal decomposition and isomerization reactions, that is, reactions that take place when a single gas is admitted to a hot vessel. After some time it became clear that although most of these reactions had complex mechanisms involving free radical intermediates and wall reactions, there did appear to be some that were simple, having mechanisms like

$$A \rightarrow B$$

or

$$A \rightarrow B + C$$

corresponding to their chemical equations. In addition, some reactions that were complex overall did appear to have simple unimolecular pathways as parts of their mechanisms. The thermal decomposition reactions studied were of many types since each organic or inorganic molecule decomposes when heated to form its own particular decomposition products. It was found that most of the characterizable unimolecular reactions are either radical elimination reactions, in which a single bond of the decomposing molecule ruptures to give two radicals, or molecular elimination reactions, in which the products are two stable molecules. The isomerizations were of the *cis–trans* type, racemizations, or ring openings to form olefins. A selection of Arrhenius parameters for the first order rate constants found for some reactions that appear to have simple mechanisms is presented in Table 5-1.

It is informative to study unimolecular gas reactions—and also termolecular gas reactions—using an experimental mixture composed of a small percentage of reacting gas diluted with a large percentage of an inert gas. In this way one ensures that the number of encounters involving one reactant molecule and one inert molecule is much larger than the number of encounters

TABLE 5-1 ARRHENIUS PARAMETERS FOR k_∞ OF SOME THERMAL UNI-MOLECULAR REACTIONS

Reactant	Product(s)	$\log_{10} A$ (sec^{-1})	E_A (kcal)
cis-2-butene	trans-2-butene	13.8	63
trans-CHD=CHD	cis-CHD=CHD	12.5	61
trans-CHCl=CHCl	cis-CHCl=CHCl	12.7	42
cis-CHϕ=CHϕ	trans-CHϕ=CHϕ	12.8	43
cis-CHϕ=CHCN	trans-CHϕ=CHCN	11.6	47
CH_3NC	CH_3CN	13.3	38
cyclopropane	propylene	15.2	65
methylcyclopropane	butene	15.4	65
1,1-dimethylcyclopropane	pentene	15.0	63
t-Butylchloride	HCl + Sec-Butene	13.9	46
4-vinyl cyclohexene	2 1,3-butadiene	15.2	62
N_2O_5	$NO_2 + NO_3$	13.7	23
ϕ–CH_3	ϕ–CH_2 + H	13.3	78
CH_3N=NCH_3	$N_2 + 2CH_3$	15.7	51
C_2H_5N=NC_2H_5	$N_2 + 2C_2H_5$	15.7	49
$Hg(CH_3)_2$	$Hg + 2CH_3$	13.3	51
$Hg(C_2H_5)_2$	$Hg + 2C_2H_5$	14.1	43
$Hg(n\text{-}C_3H_7)_2$	$Hg + 2n\text{-}C_3H_7$	15.2	46
ethane	$2CH_3$	16.5	88
CO_2	CO + O	11.3	110
CS_2	CS + S	12.9	89
COS	CO + S	11.6	68
N_2O	$N_2 + O$	11.2	60

involving two reactant molecules. For the theoretical interpretation it is then assumed that all encounters are of the former kind. We refer to inert gases added for this purpose as *diluents*.

The accumulation of experimental evidence that first order gas reactions with simple mechanisms existed was at first an awkward situation for the theory of gas reactions. The decomposing or isomerizing molecules are stable at low temperatures and react after being heated; therefore, the reaction results from increasing the energy of the molecules. From the beginning it was assumed that the activation energies measured in these reactions were equal, or almost equal, to the bond dissociation energies of the bonds that were broken. Energy transfer from the wall to the gas, however, occurs as a result of bimolecular encounters that maintain the Maxwell–Boltzmann energy distributions characteristic of the temperature of the thermostat containing the reaction vessel. Bimolecular events normally imply second order kinetics:

how may bimolecular energy transfer be reconciled with first order reaction kinetics?

The answer to this question, and a prescription for testing the answer by experiment, is that the basic mechanism—called after its originators the Lindemann–Hinshelwood mechanism—of a first order gas reaction is

$$A + M \xrightarrow{k_e} A^* + M \tag{1}$$

$$M + A^* \xrightarrow{k_{de}} A + M \tag{-1}$$

$$A^* \xrightarrow{k_{uni}} \text{products} \tag{2}$$

where A^* represents an A molecule that is *energized* in some manner as a result of the bimolecular encounter (1), and M represents any molecule that A molecules may encounter in the reaction system.

(In this chapter we depart from our usual practice of writing in the temperature dependence of rate constants. Some of the rate constants we use later refer to molecules with specific energy content: these we shall explicitly write as functions of energy, for example, $k(\varepsilon)$. Thus rate constants whose functional dependence is not stated are our customary $k(T)$'s; k^1, to be introduced shortly, is a function of pressure as well as temperature.)

If A has been diluted with a large amount of diluent gas, then M can be assumed to represent diluent gas molecules. We can envisage the "energizing" to be an increase in the A molecule's total vibrational energy such that the A^* molecule formed is capable of decomposing or isomerizing unimolecularly after encounter (1). It does not necessarily undergo a unimolecular reaction, however, for a second bimolecular event (-1) can *deenergize* the A^* molecule by removing an amount of vibrational energy sufficient to prevent unimolecular reaction. Only those A^* molecules which isomerize or decompose before a deenergizing collision occurs will form products in step (2). Making a steady state assumption on $[A^*]$ leads to the rate law

$$R \equiv \frac{d\,[\text{products}]}{dt} = \frac{k_{uni}\,k_e\,[A][M]}{k_{de}\,[M] + k_{uni}} = \frac{k_e\,[A][M]}{1 + k_{de}\,[M]/k_{uni}}$$

Two limiting forms of this rate law are of special interest. If $[M]$ is very low, then $k_{de}\,[M]$ may be much less than k_{uni}, in which case the *low pressure limit* law is $R_{\text{low pressure}} = k_e\,[A][M]$, which says that when the pressure is "very low," the rate of product formation is equal to the rate of A^* formation in reaction (1). This may be understood in terms of the time interval between collisions: the time available for A* to react unimolecularly before a bi-molecular deenergization occurs is inversely proportional to [M]. The operational definition of "very low pressure" obviously depends on the value of k_{uni}, which we shall find presently to depend on the number of atoms in A, that is, on the molecular complexity. The second limiting form, for $k_{de}[M] \gg k_{uni}$, is called the *high pressure limit*, with the rate law

$$R_{\text{high pressure}} = \frac{k_{uni} k_e}{k_{de}} [A] = k_\infty [A]$$

This rate law asserts that the rate of product formation at high pressure is equal to the rate constant for reaction (2) multiplied by $k_e [A] k_{de}^{-1}$, which is the equilibrium concentration of A*, since $k_e k_{de}^{-1} = K_{eq}$. At the high pressure limit, then, the conversion rate is determined by the unimolecular reaction rate, whereas the bimolecular energization rate determines the conversion rate at the low pressure limit.

First order gas reactions are thus high pressure limit cases of a mechanism involving three elementary reactions. We define the experimental rate constant k^1 for a first order gas reaction through the first order rate law $R = k^1 [A]^{1.0}$. We use symbol k^1, a function of p as well as T, for the first order rate constant calculated theoretically also.

By inspection from the limiting rate laws we have

$$k^1 = k_\infty = \frac{k_{uni} k_e}{k_{de}} \qquad (high\ pressure)$$

at the high pressure limit and

$$k^1 = k_e [M] \qquad (low\ pressure)$$

at the low pressure limit. At intermediate pressures k^1 undergoes a smooth transition from one limit form to another. One way to illustrate this graphically is shown in Figure 5-1, where the ratio k^1/k_∞ is plotted in a log–log presentation as a function of p_M. In general we are interested in the temperature dependence of k_∞ and k_e, and in the pressure dependence of k^1 at one temperature.

The transition from the high pressure constant $k^1 = k_\infty$ to the low pressure linear decrease in k^1 is called the *fall-off region*. The pressure range where this occurs depends on the identity of both A and M. In general, the fall-off region occurs at *higher pressures* for *smaller* M molecules. The data points indicated in Figure 5-1 for the isomerization of cyclopropane to propylene, a thoroughly studied gas reaction, show that for a molecule of this complexity the fall-off occurs in the pressure region 1 to 100 torr.

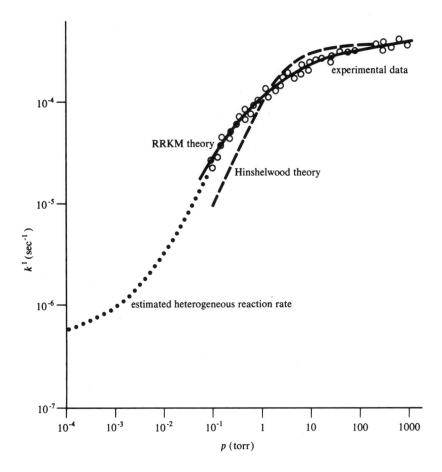

FIGURE 5-1 *Dependence of k^1 on pressure for the reaction cyclopropane $\rightarrow$ propylene at 765°K, M = cyclopropane. The experimental data are the combined results of several experimenters. The solid line represents the Hinshelwood theory, that is, $k^1 = k_\infty (1 + k_e/[M]Z_{A \cdot M})^{-1}$, with k_e calculated assuming $s = 12$. The dashed line was calculated from the RRKM theory. Note that at very low pressures k^1 again becomes constant. This is because activation occurs in collisions with the vessel wall rather than with gas molecules when the mean free path of the molecules is comparable to the dimensions of the experimental vessel. The indicated low pressure level-off has been estimated for a reaction vessel of 10 cm diameter whose walls are approximately as effective as cyclopropane in energizing cyclopropane. (For very complex molecules and reaction vessels of normal size, k_{uni} can be so small that the low pressure level-off and the high pressure limit coalesce. The probability that such molecules will decompose during a free flight from one wall collision to the next is small enough to prevent depletion of $[A^*]$ from its equilibrium value at any pressure.)*

We have made occasional references to the relationship between equilibrium constants and the ratio of forward to reverse rate constants. Unfortunately, this is of no help in relating k_e to k_{de} here, since the A* molecules, being only transient entities, cannot be characterized by thermodynamic measurements. We are left with the unfortunate situation that the predicted rate law is characterized by two parameters, namely, k_e and the ratio k_{de}/k_{uni}, and rate measurements are of no help in finding the numerator and denominator of this ratio separately. This problem is met by accepting the validity of a *strong collision assumption*. We recall from Chapter 4 that an upper limit to the rate constant of a bimolecular reaction with zero activation energy is given by the collision frequency Z. In the theory of unimolecular reactions it is assumed that all gas-kinetic A*–M collisions are "strong" in the sense that every one of them results in deenergizing A*. [In the next section we see what the strong collision assumption means for the energizing reaction (1).] If k_{de} is set equal to the collision frequency Z_{A^*M}, and if the conversion rate has been measured in both the high and low pressure limits, or in one limit and a substantial part of the transition region, then values for k^1 yield both k_e and k_{uni} in the above mechanism.

If for a given reaction graphs similar to Figure 5-1 are constructed for different diluent gases, it is found that the fall-off region shifts to higher or lower pressures for each diluent. Different M molecules have different *efficiencies* in energizing A molecules or deenergizing A* molecules, that is, different values of k_e and k_{de}. Since the ratio k_e/k_{de} corresponds to an equilibrium constant, its value is independent of the nature of M. Diluent gases that are ineffective collisional energizers or deenergizers have fall-off regions shifted to higher pressures. The relative efficiencies for a number of diluent gases in the cyclopropane → propylene reaction are presented in Table 5-2.

EXERCISE 5-1 (a) Write out in detail the argument in the preceding paragraph, deriving a formula for k^1/k_∞, and taking $k^1 = 0.5k_\infty$ as a measure for the location of the fall-off region. (b) Describe how one could determine the ratio $k_{e, M}/k_{e, A}$ where $k_{e, A}$ is k_e for M = A and $k_{e, M}$ is k_e for some diluent gas. (c) Show from the collision frequency formula how this ratio should be corrected for reduced mass and collision diameter to give the relative energy transfer efficiency on a "per collision" basis. (d) From the data in Table 5-2 and Figure 5-1 back-calculate $k_{e, He}/k_{e, A}$ and the pressure of He for which $k^1 = 0.5k_\infty$ in an experiment with $\Delta = [C_3H_6] \ll [He]$.

Table 5-2 shows that a wide range of relative efficiencies is found for the different diluents. Large polyatomic molecules have high efficiencies; monatomic gases have low efficiencies; small molecules are intermediate. This behavior is characteristic of energy transfer studied in the fall-off region of unimolecular reactions. It is not the same as the behavior found in other

TABLE 5-2   RELATIVE ENERGY TRANSFER EFFI-
CIENCIES ON A PER-COLLISION BASIS[a]

Diluent gas	$(k_{e, M}/k_{e, \Delta})(\sigma_\Delta/\sigma_M)^2(\mu_\Delta/\mu_M)^{-\frac{1}{2}}$
cyclopropane	1.0
helium	0.05
argon	0.07
nitrogen	0.07
carbon monoxide	0.08
hydrogen	0.12
methane	0.24
water	0.74
benzotrifluoride	0.75
mesitylene	0.89
toluene	1.10

[a] Derived from fall-off experiments on cyclopropane isomerization at 765°K.

experiments on bimolecular energy transfer, principally because we are dealing here with VV transfer (Section 4-6) from highly excited vibrational states. We find later that the detailed theory of unimolecular reactions suggests that the relative efficiencies of different diluents should be obtained from low pressure experiments only, since the fall-off behavior of k^1 in the region is a combination of energy transfer and reactivity effects.

Before proceeding to the theory of unimolecular reactions, let us note an important feature of Table 5-1.   The A factors associated with k_∞ range from about 10^{11} to 10^{16} sec^{-1}.   For comparison, we recall that molecular vibration frequencies are in a similar range, for instance 10^{12}–10^{14} sec^{-1}, to correspond by way of $v = c/\lambda$ to the fundamental vibrational transitions generally observed in ir spectra between 3 and 300 μ.   The coincidence is not entirely accidental, as we shall presently find.

5-2   THEORY OF UNIMOLECULAR REACTIONS

The set of three elementary reactions comprising the basic mechanism of unimolecular gas reactions is found to provide a suitable framework for a qualitative description of experimental observations made on this class of reactions. We now wish to extend our study of unimolecular reactions to the major theories that attempt to provide formulas for calculating k^1 values in quantitative agreement with experiments.   In this area of physical chemistry, theoretical research has been remarkably successful in providing not only computational tools, but also profound insight into the microscopic character of the observed phenomena.   Unfortunately, the algebra associated with much of the theory is voluminous, and we are forced to forego presenting it. We emphasize instead the ideas incorporated into the theory and the consequences of working these ideas out.

5-2a First Theory: Activation in Many Degrees of Freedom

We have already seen that if k_{de} is taken as Z_{A*M}, then two parameters, for example k_e and k_∞, are required to characterize k^1 in the rate law for the above mechanism for all pressures. On the theoretical side, then, two parameters should also be provided. These could be k_e and k_∞, or they could be k_e and k_{uni}, or k_{uni} and k_∞. The first theoretical developments, however, only attempted to provide one of them. This allowed derivation of expressions for the one-parameter ratio k^1/k_∞, which characterizes the fall-off curve if k_∞ is known. The complete modern theory we discuss later provides a means of calculating k^1 itself.

Let us begin by multiplying the k^1 formula

$$k^1 = \frac{k_e k_{uni} [M]}{k_{uni} + k_{de} [M]}$$

by k_{de}/k_{de} and factoring to obtain

$$k^1 = \left(\frac{k_e}{k_{de}}\right)\left(\frac{k_{de} [M]}{k_{uni} + k_{de} [M]}\right) k_{uni}$$

The first factor is K_{eq} for the energizing–deenergizing reaction. Since $[A^*] \ll [A]$ for all conditions, $K_{eq} = [A^*]/[A]$ is almost exactly equal to $[A^*]/([A] + [A^*])$, which is the probability P that an A molecule selected at random from a system in which the $A \rightleftarrows A^*$ equilibrium is maintained has sufficient energy to belong to the A^* category. In the second factor we replace k_{de} with the collision frequency Z (Section 4-2) on the strong collision assumption mentioned in the previous section. Then the k^1 equation becomes

$$k^1 = P\left(\frac{Z [M]}{k_{uni} + Z [M]}\right) k_{uni}$$

and the rate law is

$$R = P [A]\left(\frac{Z [M]}{k_{uni} + Z [M]}\right) k_{uni}$$

In this formulation the consequences of the three-reaction mechanism become very clear. The rate of product formation is equal to the equilibrium concentration of A*, given by P [A], multiplied by a factor Z [M] $(k_{uni} + Z$ [M]$)^{-1}$, which corrects the equilibrium concentration of A* for the low or intermediate pressure ranges in which [A*] falls below its equilibrium value owing to the failure of the energizing–deenergizing reactions to maintain equilibrium against the depleting effect of the unimolecular reaction, and multiplied again by the rate constant k_{uni} for reaction of A* to form products. The fall-off in k^1 from its high pressure limit value k_∞ is therefore due to a decreasing value of [A*].

At the high pressure limit, the factor Z [M]$(k_{uni} + Z$[M]$)^{-1}$ becomes unity, and the rate law assumes its high pressure limiting form

$$R = Pk_{uni} [A]$$

$$= k_\infty [A]$$

The high pressure rate constant k_∞ is thus composed of the probability factor P and the unimolecular rate constant factor k_{uni}. Referring to the collection of experimental Arrhenius parameters for k_∞ in Table 5-1, we can appreciate the working assumption of the first researchers that the A factors represented k_{uni} values and the $\exp(-E_A/RT)$ represented P values. This appears quite plausible since $\exp(-E_A/RT)$ looks like a Boltzmann factor in the energy required for forming an A* and since the A factors resemble molecular vibration frequencies.

This interpretation proves to be false. If $P = \exp(-E_A/RT)$, then since $P = K_{eq} = k_e/k_{de} = k_e/Z$, the energizing reaction rate constant k_e would be $k_e = Z \exp(-E_A/RT)$. This is the line-of-centers rate constant formula, which in itself is plausible. When experimental values of k_e were measured, however, they proved to be *much larger* than $Z \exp(-E_A/RT)$.

This presented the theory with a twofold problem. First, if k_e is much larger than $Z \exp(-E_A/RT)$, then P is much larger than $\exp(-E_A/RT)$. Second, experimental A factors do not equal k_{uni}; actually k_{uni} is much smaller than A. If we anticipate the correct interpretation of these problems, the high values for k_e and P arise because energization is an activation-in-many-degrees-of-freedom problem, and the k_{uni} values are much less than molecular vibration frequencies because an energized molecule does not react until the energy necessary for reaction is concentrated in the particular part of the molecule where reaction occurs.

The first successful effort (attributed to Hinshelwood in 1926) to derive a theoretical fall-off curve in reasonable agreement with experiment involved replacing $k_e = Z \exp(-E_A/RT)$ with the activation-in-many-degrees-of-freedom expression derived in Section 4-3:

$$k_e = Z \, \Gamma(s)^{-1} \, [\varepsilon_0/kT]^{s-1} \exp(-\varepsilon_0/kT)$$

where s is the number of oscillators contributing to energization, and ε_0 is the *critical energy* that a molecule must have before unimolecular reaction is possible. In unimolecular reactions the energy is all (or almost all, if rotations can also contribute to ε_0) in molecular vibrations; in the theory of unimolecular reactions it is customary to speak of a vibrationally energized molecule as a collection of s *effective oscillators* that will undergo unimolecular reaction when the energy ε_0 is concentrated in the reactive part of the molecule. Since $[\Gamma(s)]^{-1}[\varepsilon_0/kT]^{s-1}$ can be made much greater than unity for $s > 1$ and $\varepsilon_0 > (s-1)kT$, the usual case in experiments, k_e can be adjusted upwards as desired by selecting the appropriate value of s.

To achieve agreement between theoretical and experimental k_e values (cf. Figure 5-1) it was found in practice that s generally had to be set equal to about one half the total number of vibrations ($3N - 6$ for a nonlinear N-atomic molecule, $3N - 5$ for a linear N-atomic molecule), the critical energy ε_0 in ergs being set equal to the experimental activation energy E_A for k_∞ in kilocalories divided by 1.44×10^{13}.

EXERCISE 5-2 (a) Calculate the ratios k_e/k_{de} at $1,000°K$ for a reaction with $E_4 = 40$ kcal, for $s = 1, 4, 12$, and 20. (b) Show that the value of [M] when $k^1 = 0.5$ k_∞ is $k_\infty k_e^{-1}$. (c) Calculate the pressure of M when $k^1 = 0.5 k_\infty$ for $s = 1, 4, 12$, and 20 if $k_\infty = 10^7$ sec^{-1}.

EXERCISE 5-3 (a) Show by rearranging the rate law for a unimolecular reaction that a graph of $1/k^1$ versus $1/p_M$ should be a straight line. What do the slope and intercept of this graph represent? (b) Calculate the slopes and intercepts predicted by the collision theory formula for $s = 1$ and $s = 4$ for the conditions of Exercise 5-2.

EXERCISE 5-4 Activation energies E_A for unimolecular reactions are measured for k_∞ and then associated with the critical energy ε_0. For $s = 4$ and the conditions of Exercise 5-2, what activation energy $E_A \equiv R(d \ln k^1/dT)$ would be measured at the low pressure limit? Compare Exercise 3-6.

Although this first-generation theory yields reasonable fall-off graphs, it has shortcomings. First, the effective number of oscillators s is not predicted by the theory but left as a disposable parameter that can be adjusted to force theory and experiment into agreement. Second, and more important, this theory does not consider the dependence of k_{uni} on energy in excess of the critical amount ε_0. Although it appears reasonable that all vibrational degrees of freedom might not contribute to the critical oscillator, it does not appear reasonable that one could really obtain a correct theory without taking account of the more rapid unimolecular reaction that is likely to result when ε is much greater than ε_0. The second-generation theory, which we consider next, attempts to calculate the fall-off graph by explicitly taking this into account. The parameter s still remains unspecified, however.

5-2b   RRK Theory: Energy-Dependent Unimolecular Reaction Rate

We consider now that a molecule is a collection of s oscillators coupled to one another in such a way that energy is periodically exchanged among them at a frequency ω sec^{-1}. We calculate a fall-off graph by finding k_{uni} as a function of ε and either integrating this k_{uni} function over all ε for a classical mechanical formula or summing over oscillator energy eigenvalues for a quantum mechanical formula. The results for the two procedures are closely equivalent; the classical approach is used here. It is called the RRK theory after its inventors Rice, Ramsperger, and Kassel.

We require two postulates. The first one has the same name as, but a wider meaning than, the strong collision assumption of the previous theory. The *strong collision assumption* now asserts that the distribution of oscillator energies resulting from an energizing or deenergizing collision is *random*; this means that statistical redistributions of thermal energy between the set of s oscillators and a heat bath composed of M molecules occur Z [M] times per second. The random character of the redistributions allows use of statistical methods to discover what these distributions are.

(Earlier we used the term strong collision assumption to mean that deenergization of A* was assumed to occur at every collision with M. The present version of the strong collision assumption implies the former version also, since the statistical redistribution is found to lower the energy of virtually all A* molecules below ε_0. The term "random incidence assumption" might be a preferable designation for our new strong collision assumption.)

Our second postulate is called the *free energy migration* assumption. It asserts that reallocation of vibrational energy *within* the reacting molecule is *also random*; this implies another statistical redistribution, but of the specific energy ε rather than of thermal energy, among the s oscillators at frequency ω. Statistical methods can be used to discover these distributions also.

The probability in classical statistical mechanics that a particular one of the oscillators, which we call the *critical oscillator*, has energy in excess of ε_0 when the total energy of all the s oscillators is ε is given by the formula

$$\left(\frac{\varepsilon - \varepsilon_0}{\varepsilon}\right)^{s-1}$$

In deriving this formula it is assumed that ε and ε_0 are large and that ε is greater than ε_0. Since reaction is assumed to occur whenever the critical oscillator has energy greater than ε_0, the energy-dependent k_{uni} is

$$k_{uni}(\varepsilon) = \omega\left(\frac{\varepsilon - \varepsilon_0}{\varepsilon}\right)^{s-1}$$

To go with our energy dependent k_{uni} we need an energy-dependent P. The probability in classical statistical mechanics that a given set of s oscillators has total energy in the range ε to $\varepsilon + d\varepsilon$, when the set of oscillators is in equilibrium with a thermostat at temperature T, is given by the formula

$$P(\varepsilon)\,d\varepsilon = \frac{\varepsilon^{s-1}e^{-\varepsilon/kT}}{\Gamma(s)(kT)^s}\,d\varepsilon$$

We showed earlier that $k_\infty = Pk_{uni}$. The present theory then has

$$k_\infty = \int_{\varepsilon_0}^\infty \frac{\omega(\varepsilon - \varepsilon_0)^{s-1}e^{-\varepsilon/kT}}{\Gamma(s)(kT)^s}\,d\varepsilon$$

The lower limit of the integral is ε_0 rather than zero because $k_\infty = 0$ if ε is less than ε_0. We are primarily interested in the fall-off graph, however. We obtain k^1/k_∞ by writing the factored form $k^1 = PZ\,[M]\,(k_{uni} + Z\,[M])^{-1} \times k_{uni}$ as a function of energy

$$k^1(\varepsilon) = P(\varepsilon)Z\,[M]\,(k_{uni}(\varepsilon) + Z\,[M])^{-1}\,k_{uni}(\varepsilon)$$

and integrating from $\varepsilon = \varepsilon_0$ to $\varepsilon = \infty$

$$k^1 = \int_{\varepsilon_0}^\infty k^1(\varepsilon)\,d\varepsilon = \int_{\varepsilon_0}^\infty \frac{P(\varepsilon)Z\,[M]\,k_{uni}(\varepsilon)}{k_{uni}(\varepsilon) + Z\,[M]}\,d\varepsilon$$

$$= \int_{\varepsilon_0}^\infty \frac{e^{-\varepsilon/kT}Z\,[M]\,\omega(\varepsilon - \varepsilon_0)^{s-1}}{\Gamma(s)(kT)^s\left(\omega\left(\dfrac{\varepsilon - \varepsilon_0}{\varepsilon}\right)^{s-1} + Z\,[M]\right)}\,d\varepsilon$$

Introducing the substitution

$$x = (\varepsilon - \varepsilon_0)/kT$$

and simplifying yields

$$k^1 = \frac{\omega e^{-\varepsilon_0/kT}}{\Gamma(s)} \int_0^\infty \frac{x^{s-1}e^{-x}\,dx}{1 + \dfrac{\omega}{Z\,[M]}\left(\dfrac{x}{x + \varepsilon_0/kT}\right)^{s-1}}$$

In the high pressure limit the second term in the denominator of the integral vanishes, and the integral becomes identical to the definition of $\Gamma(s)$. Thus, in this theory

$$k_\infty = \omega \exp(-\varepsilon_0/kT)$$

The fall-off graph is therefore given in RRK theory by

$$\frac{k^1}{k_\infty} = \Gamma(s)^{-1} \int_0^\infty \frac{x^{s-1} e^{-x} \, dx}{1 + \dfrac{\omega}{Z\,[M]} \left(\dfrac{x}{x + \varepsilon_0/kT}\right)^{s-1}}$$

A quantum statistical version of this theory, in which the oscillator energies must be multiples of $h\nu_0$, yields a summation here rather than an integral.

EXERCISE 5-5   As an alternative derivation of the k_∞ equation, show that s-fold repeated application of the differentiation by parts formula will lead from

$$k_\infty = \int_0^\infty \omega\, (\varepsilon - \varepsilon_0)^{s-1} \exp(-\varepsilon/kT)\, \Gamma(s)^{-1}(kT)^{-s} \, d\varepsilon$$

to $k_\infty = \omega \exp(-\varepsilon_0/kT)$ for (positive) integral values of s.

EXERCISE 5-6   At the high pressure limit, the assumptions of the activated complex theory can be assumed to be closely fulfilled.   (a) Write out the ACT formula for k_∞.   (b) How is ω interpreted in this theory?   (c) Describe possible activated complexes that would account for $\omega \gg kT/h$.   (d) How would $\omega \ll kT/h$ be interpreted in ACT?   (e) Find plausible examples for (c) and (d) in Table 5-1.

EXERCISE 5-7   How many ways are there to put n peppercorns into a pair of slippers?   Of these ways, how many will involve putting at least m peppercorns ($m < n$) into the left slipper?   What is the probability of finding at least m of the n peppercorns in the left slipper if you know that the distribution of peppercorns into slippers was made randomly?   What correction should be made to the formula $P = (\varepsilon - \varepsilon_0/\varepsilon)^{s-1}$ for $s = 2$?

EXERCISE 5-8   Show that the low pressure limit of the above integral equation k^1/k_∞ leads by repeated application of the integration by parts formula to the activation-in-many-degrees-of-freedom formula for k_e for the case $\varepsilon_0 \gg kT$.

Since ω and ε_0/kT are known from the Arrhenius parameters of k_∞, the "RRK integral," which must be evaluated by a numerical integration, gives the fall-off graph corresponding to all values of s.   The values of s that give best agreement with experimental fall-off data again turn out to be about one half the total number of vibrational degrees of freedom of the reacting molecule, as in the activation-in-many-degrees-of-freedom theory for k_e

described earlier. Now, however, we have taken explicit account of the energy dependence of k_{uni}, and it turns out that the resulting integral equation for the fall-off graph gives moderate improvement in the agreement between calculated and observed fall-off graphs. What we have not yet done is to calculate k_{uni} from the properties of the reacting molecules, for ω is still a purely experimental quantity.   Moreover, s is also an adjustable parameter that can be varied, within limits, to obtain agreement with experiment.

These shortcomings are absent from the theory we consider next, which represents the present state of the art in theoretical understanding of unimolecular reactions.

5-2c   RRKM Theory: Complete Calculation of k[1]

The RRKM theory, whose letters stand for Rice, Ramsperger, Kassel, and Marcus, combines most of the formalism of the preceding theory with activated complex ideas to allow computation of k_{uni} from properties of the reacting molecules.   The critical energy ε_0 is still derived from the experimental E_A (as was E_b in the bimolecular ACT calculations discussed in Section 4-7); a degree of arbitrariness in choice of activated complex, which for unimolecular reactions is termed the *critical configuration*, also remains.   Within these limitations, disagreement between the RRKM theory, or computationally modified forms of it, and experiment can only be due to failure of the strong collision assumption or of the *random lifetime assumption*, which we now introduce as an alternative to the periodic random reallocation of vibrational energy postulated in the free energy migration assumption of the RRK theory.

The random lifetime assumption asserts that A molecules that become energized A_ε^* molecules (the subscript denotes vibrational energy in the range ε to $\varepsilon + d\varepsilon$) capable of reacting unimolecularly will have *lifetimes* τ, defined as the time interval between collisional energization and formation of product molecules, which are distributed randomly among the many quantum states with energy in the range ε to $\varepsilon + d\varepsilon$.   An immediate analogy would be to the lifetimes of the atoms in a sample of a radioactive isotope, which also have a random distribution.   The probability that an atom of a radioactive isotope with a first order decay constant k will have a lifetime in the range from τ to $\tau + d\tau$ is given by the probability density function $P(\tau) = k \exp(-k\tau)$. Similarly, the probability density function for an A_ε^* molecule having a lifetime in the range τ to $\tau + d\tau$ is

$$P(\tau, \varepsilon) = k_a(\varepsilon) \exp[-k_a(\varepsilon)\tau]$$

where $k_a(\varepsilon)$ is the energy-dependent unimolecular decay constant that will be our task to evaluate by ACT methods.

EXERCISE 5-9 Construct a logical argument that demonstrates the validity of $P(\tau) = k \exp(-k\tau)$ for radioactive decay of A atoms. You might find a graph of log [A] versus time useful.

The random lifetime assumption clears the path for a complete calculation of k^1 using ACT methods *without* invoking the ACT postulate that the activated complex concentration is given by consideration of a reactant $\rightleftarrows$ activated complex equilibrium. If the random lifetime assumption is true, then the ACT result for the decomposition rate of the critical configuration will be valid even at low pressures, where it is certainly *not* true that an equilibrium exists between energized A^* molecules and critical configuration A^+ molecules until the over-all reaction $A \rightleftarrows$ products is at equilibrium.

The derivation begins again with the strong collision assumption, asserting that A molecules are reallocated into quantum states at the collision rate $Z\,[M]$. A_ε molecules are thus again assumed to be generated at the rate $Z\,[M][A]\,P(\varepsilon)d\varepsilon$ where $P(\varepsilon)d\varepsilon$ is the equilibrium probability of an A molecule having energy in the range ε to $\varepsilon + d\varepsilon$. The function $P(\varepsilon)$ we again presume to be calculable by the methods of statistical mechanics. If an A_ε molecule has energy greater than the critical energy ε_0, it is an A_ε^* molecule and may either react unimolecularly or be deenergized to an A_ε with ε less than ε_0. The strong collision assumption again implies that the deenergization rate is $Z\,[M][A_\varepsilon^*]$. Since deenergizing collisions are randomly distributed in time, the probability of unimolecular reaction, which is equal to the probability of an A^* molecule *not* being deenergized within its lifetime τ, is given by

$$P_{\text{reaction}}(\tau) = \exp(-Z\,[M]\,\tau)$$

The validity of this equation can be seen by repeating the argument used in Exercise 5-9. The rate of producing A_ε^* molecules that have lifetimes between τ and $\tau + d\tau$ and react within their lifetime (i.e., the reaction rate for energy in the range from ε to $\varepsilon + d\varepsilon$) is then the product of three factors: the rate of generating A_ε^* molecules $(Z\,[M][A]\,P(\varepsilon)\,d\varepsilon)$, the probability of having lifetime between τ and $\tau + d\tau\,(P(\tau, \varepsilon)\,d\tau)$, and the probability of reaction before collision $(P_{\text{reaction}}(\tau))$:

$$dR_{\varepsilon,\,\tau} = Z\,[M][A]\,P(\varepsilon)P(\tau, \varepsilon)P_{\text{reaction}}(\tau)\,d\varepsilon\,d\tau$$

The rate constant $k(\varepsilon)$ is obtained by integrating over τ and dividing by [A]

$$k(\varepsilon)\,d\varepsilon = Z\,[M]\,P(\varepsilon)\int_0^\infty k_a(\varepsilon)\exp[-(k_a(\varepsilon) + Z\,[M])\tau]\,d\tau\,d\varepsilon$$

$$= \frac{k_a(\varepsilon)P(\varepsilon)\,d\varepsilon}{1 + k_a(\varepsilon)/Z\,[M]}$$

The first order rate constant k^1 is obtained by integrating $k(\varepsilon)\, d\varepsilon$ from ε_0 to ∞:

$$k^1 = \int_{\varepsilon_0}^{\infty} \frac{k_a(\varepsilon)P(\varepsilon)\, d\varepsilon}{1 + k_a(\varepsilon)/Z\,[M]}$$

We are now back in much the same position as we were with the previous theory as far as the form of the k^1 integral is concerned; that is, our present $k_a(\varepsilon)$ looks just like k_{uni} in the RRK theory. $P(\varepsilon)$ is in principle the same function as in the RRK theory, although we will use a quantum statistical formula to evaluate it. The great contrast to our former position, however, is that we are now able, by the random lifetime assumption, to calculate $k_a(\varepsilon)$ rather than infer it from experiments.

Before proceeding to find $k_a(\varepsilon)$, let us take care of $P(\varepsilon)$. The quantum statistical formula for the probability of any system with quantum state density $g(\varepsilon)$ having energy in the range ε to $\varepsilon + d\varepsilon$ is

$$P(\varepsilon)\, d\varepsilon = g(\varepsilon)\exp(-\varepsilon/kT)q_v^{-1}\, d\varepsilon$$

where we denote the partition function by q_v (cf. Section 4-7) since we are now interested only in the distribution in vibrational quantum states. (Contributions to ε from rotational degrees of freedom are also possible (*v. infra*); in this case q_v would be multiplied by the rotational partition functions for the contributing rotational degrees of freedom. A correction to take molecular symmetry into account is also sometimes needed.) Since the integration of $k_a(\varepsilon)\, d\varepsilon$ starts at $\varepsilon = \varepsilon_0$, $g(\varepsilon)$ is actually needed only for energized molecules; that is, we require only $g^*(\varepsilon)$. If k_∞ only is to be calculated, $g^*(\varepsilon)$ divides out of the final equation; otherwise it is required explicitly. Substituting $P(\varepsilon)$ into k^1 yields

$$k^1 = \frac{1}{q_v}\int_{\varepsilon_0}^{\infty} \frac{k_a(\varepsilon)g^*(\varepsilon)e^{-\varepsilon/kT}\, d\varepsilon}{1 + k_a(\varepsilon)/Z\,[M]}$$

To accomplish the $k_a(\varepsilon)$ evaluation, we note first that at very high pressure there would be an equilibrium $A_\varepsilon^* \rightleftharpoons A_\varepsilon^+$ between energized molecules that merely have ε in the range ε to $d\varepsilon$, the A_ε^*, and those molecules in the same energy range that are in the critical configuration A_ε^+. Since these have the *same energy*, by the fundamental rules of statistical mechanics the ratio of their concentrations is given by the ratio of the number of critical configuration quantum states with energy between ε and $\varepsilon + d\varepsilon$, $g^+(\varepsilon)$, to the total number of quantum states in the same interval, $g^*(\varepsilon)$. Mathematically, $[A_\varepsilon^+]/[A_\varepsilon^*] = g^+(\varepsilon)/g^*(\varepsilon)$. The rate of reaction for molecules in the energy range ε to $\varepsilon + d\varepsilon$, $k_a(\varepsilon)[A_\varepsilon^*]\,d\varepsilon$, is then expressible in terms of the concentration of critical configuration molecules in the same energy range as $k^+(\varepsilon)[A_\varepsilon^+]\,d\varepsilon$. Thus $k_a(\varepsilon)$ for infinite pressure is given by

$$k_a(\varepsilon) = \frac{[A_\varepsilon^+]}{[A_\varepsilon^*]}\,k^+(\varepsilon) = \frac{g^+(\varepsilon)}{g^*(\varepsilon)}\,k^+(\varepsilon)$$

We shall presently find that $k^+(\varepsilon)$, the average rate constant for reaction of critical configuration molecules, can be obtained as a simple one-dimensional ACT problem. *By the random lifetime assumption the above equation for $k_a(\varepsilon)$ is valid not just at infinite pressure where the equilibrium $A_\varepsilon^* \rightleftarrows A_\varepsilon^+$ is always maintained by collisions, but at any pressure.*

One degree of freedom of the critical configuration, called the *critical coordinate*, is assumed to be separable from the other degrees of freedom and to behave as translational motion over a small distance δ. Only part of the energy ε of an A_ε^+ molecule will be translational energy along the critical coordinate. This translational energy, which we call ε_t, will be different from one A_ε^+ molecule to another. It is given by the usual kinetic energy formula $\varepsilon_t = \frac{1}{2}\mu v^2$, where μ is the effective mass associated with translational motion along the critical coordinate, and the corresponding velocity along the critical coordinate is $v = (2\varepsilon_t/\mu)^{\frac{1}{2}}$. The minimum value of ε_t is zero. Since the ε_0 part of the total A^+ energy ε represents a potential energy needed to form A^+ from A, the maximum value of ε_t is $\varepsilon - \varepsilon_0$.

The rate constant for crossing the small length δ of the critical coordinate at this velocity in the direction corresponding to the reaction $A_\varepsilon^+ \rightarrow$ products is

$$k^+(\varepsilon_t) = v/2\delta = (\varepsilon_t/2\mu)^{\frac{1}{2}}\delta^{-1}$$

where the factor $\frac{1}{2}$ accounts for the fact that ε_t may correspond with equal probability to motion in either direction along the critical coordinate. A second equation for δ can be derived from the quantum mechanical energy level formula for the one-dimensional particle-in-a-box problem

$$\varepsilon_t = \frac{h^2}{8\mu}\frac{n_x^2}{\delta^2}$$

$$n_x = \left(\frac{8\mu\delta^2\varepsilon_t}{h^2}\right)^{\frac{1}{2}}$$

$$\frac{dn_x}{d\varepsilon_t} \equiv g^+(\varepsilon_t) = \left(\frac{2\mu\delta^2}{h^2\varepsilon_t}\right)^{\frac{1}{2}}$$

$$\delta = g^+(\varepsilon_t)\left(\frac{h^2\varepsilon_t}{2\mu}\right)^{\frac{1}{2}}$$

The function $g^+(\varepsilon_t)$, called the quantum state density for the critical coordinate, should not be confused with $g^+(\varepsilon)$. Substitution of this equation for δ into the equation for $k^+(\varepsilon_t)$ gives as the rate constant for reaction of A_ε^+ molecules with critical coordinate energy in the range ε_t to $\varepsilon_t + d\varepsilon_t$

$$k^+(\varepsilon_t) = h^{-1}g^+(\varepsilon_t)^{-1}$$

The number of quantum states of critical configuration molecules within a translational energy increment $d\varepsilon_t$ is the product of the quantum state density for the critical coordinate $g^+(\varepsilon_t)$ and the quantum state density for all other degrees of freedom of the critical configuration molecule with energy $\varepsilon - \varepsilon_t$, which we call $g^+(\varepsilon - \varepsilon_t)$. If the product $g^+(\varepsilon_t)g^+(\varepsilon - \varepsilon_t)$ is multiplied by the rate constant for the same energy range, $k^+(\varepsilon_t)$, and integrated over the possible range of ε_t, which is from zero to $\varepsilon - \varepsilon_0$, then we obtain the average rate constant $k^+(\varepsilon)$ multiplied by the total number of critical configuration quantum states between ε and $\varepsilon + d\varepsilon$, which is $g^+(\varepsilon)$:

$$g^+(\varepsilon)k^+(\varepsilon) = \int_0^{\varepsilon - \varepsilon_0} h^{-1}g^+(\varepsilon_t)^{-1}g^+(\varepsilon_t)g^+(\varepsilon - \varepsilon_t)\, d\varepsilon_t$$

$$= h^{-1} \int_0^{\varepsilon - \varepsilon_0} g^+(\varepsilon - \varepsilon_t)\, d\varepsilon_t$$

To complete the derivation we return to our original equation for $k_a(\varepsilon)$, namely, $k_a(\varepsilon) = g^+(\varepsilon)k^+(\varepsilon)/g^*(\varepsilon)$, substitute the preceding equation for $g^+(\varepsilon)k^+(\varepsilon)$, and obtain for our general $k_a(\varepsilon)$ equation

$$k_a(\varepsilon) = \frac{1}{hg^*(\varepsilon)} \int_0^{\varepsilon - \varepsilon_0} g^+(\varepsilon - \varepsilon_t)\, d\varepsilon_t = \frac{N^+(\varepsilon - \varepsilon_0)}{hg^*(\varepsilon)}$$

At this point we have finished the formal part of the RRKM theory: this $k_a(\varepsilon)$ equation can be substituted into the integral appearing in the equation for k^1; the combination can then be integrated over all ε from ε_0 to ∞ to fulfill our initial goal of deriving a formula for k^1 as a function of [M] and T.

Reducing the formal k^1 equation to a computational equation for a particular reaction requires substantial additional effort. We must have formulas for Z, q_v, $g^*(\varepsilon)$, and $N^+(\varepsilon - \varepsilon_0)$. The encounter frequency Z in cm^3 sec^{-1} is readily obtained as a function of temperature by inserting appropriate values of σ_{AM} and μ_{AM} into (cf. Section 4-2)

$$Z = \sigma_{AM}^2(8\pi kT/\mu_{AM})^{\frac{1}{2}}$$

The factor $Z\,[M]$ is conveniently written for computational purposes as a function of pressure

$$Z\,[M] = 4.41 \times 10^7 \sigma_{AM}^2 T^{-\frac{1}{2}} \mu_{AM}^{-\frac{1}{2}}\, p_M$$

where Z [M] is in inverse seconds, σ_{AM} is in angstroms, T is in degrees K, and p_M is in torr.

The real computational effort necessary to provide all the factors in the equation for k^1 thus resolves itself into a threefold requirement: we need formulas for q_v, $g^*(\varepsilon)$, and $N^+(\varepsilon - \varepsilon_0)$. The partition function q_v for the A molecules is evaluated by conventional methods discussed in statistical mechanics textbooks and mentioned before in connection with ACT calculations in Section 4-7. Evaluation of $g^*(\varepsilon)$ and $N^+(\varepsilon - \varepsilon_0)$ can be done in a great variety of ways, depending on the problem under investigation and the personality of the investigator.

One possibility is to use the classical formulas employed in the RRK theory but corrected for zero-point energy to give *semiclassical* formulas of improved accuracy. These formulas are

$$g^*(\varepsilon) = \frac{(\varepsilon + \varepsilon_z^*)^{s-1}}{\Gamma(s) \prod\limits_{i=1}^{s} h v_i^*}$$

$$N^+(\varepsilon - \varepsilon_0) = \frac{(\varepsilon - \varepsilon_0 + \varepsilon_z^+)^{s-1}}{\Gamma(s) \prod\limits_{i=1}^{s-1} h v_i^+}$$

where $\varepsilon_z^* \equiv \frac{1}{2} \sum\limits^{s} h v_i^*$ and $\varepsilon_z^+ \equiv \frac{1}{2} \sum\limits^{s-1} h v_i^+$ are the zero-point energies of A* and A$^+$, $\{v_i^*\}$ and $\{v_i^+\}$ are the fundamental vibration frequencies of A* and A$^+$, and s refers to *all* the vibrations of A*. [The formula for $N^+(\varepsilon - \varepsilon_0)$ is obtained by integrating the $g^+(\varepsilon)$ formula, replacing ε by $\varepsilon - \varepsilon_0$ to account for the different energy zeros for A$^+$ and A*, and replacing s by $s - 1$ to account for the fact that one of the vibrations of A* becomes the critical coordinate of A$^+$.] Further improvement on the accuracy of the $g^*(\varepsilon)$ formula is obtained by multiplying ε_z^* by a correction factor a; empirical methods for estimating a have been devised. A correction can be made to the ratio $g^+(\varepsilon)/g^*(\varepsilon)$ in the original $k_a(\varepsilon)$ formula to take into account the effect of rotation; a common procedure is to multiply $g^+(\varepsilon)/g^*(\varepsilon)$ by the ratio of the rotational partition functions $q_r^+/q_r^* = (I_x^+ I_y^+ I_z^+/I_x^* I_y^* I_z^*)^{\frac{1}{2}}$, where the I's are the principal moments of inertia. This allows for the effect that the different sizes of A* and A$^+$ will have upon the density of quantum states ratio. Combining our formulas gives for *the semiclassical $k_a(\varepsilon)$ formula*

$$k_a(\varepsilon) = \frac{q_r^+}{q_r^*} \left(\frac{\varepsilon - \varepsilon_0 + \varepsilon_z^+}{\varepsilon + \varepsilon_z^*} \right)^{s-1} \left(\prod_{i=1}^{s} v_i^* \Big/ \prod_{i=1}^{s-1} v_i^+ \right)$$

If a set of vibration frequencies for the energized molecules and the critical configuration is assumed, this formula becomes an explicit function of ε

that can be inserted into the k^1 integral equation for calculation of k^1 as a function of pressure.   This $k_a(\varepsilon)$ formula is in fact exactly the RRK formula of the previous section, $k_{un\,i}(\varepsilon) = \omega(\varepsilon - \varepsilon_0/\varepsilon)^{s-1}$, corrected for zero-point energy and having the energy reallocation frequency ω given by

$$\omega = \left(q_r^+ \prod_{i=1}^{s} v_i^* \Big/ q_r^* \prod_{i=1}^{s-1} v^+ \right)$$

EXERCISE 5-10   (a) Show that the RRKM formula for k^1 can be written

$$k^1 = \frac{kT}{h} \frac{q_r^+}{q_r^*} \frac{e^{-\varepsilon_0/kT}}{q_v} \int_0^\infty \frac{\int_0^{\varepsilon^*} g^+ \, (\varepsilon^* - \varepsilon_t) \, d\varepsilon_t e^{-\varepsilon^*/kT} \, d(\varepsilon^*/kT)}{1 + k_a(\varepsilon^*)/Z\,[M]}$$

where $\varepsilon^* = \varepsilon - \varepsilon_0$.   Be careful with the change of variables and note the change in meaning between $g^+(\varepsilon - \varepsilon_t)$ and $g^+(\varepsilon^* - \varepsilon_t)$.   (b) Demonstrate the correlation between the k_∞ from this form of the RRKM equation for k^1 and the formula for k_∞ that one would derive directly from the activated complex theory.

If the semiclassical quantum state density formulas are accepted, then all we have done with the RRKM theory amounts to an identification of and means of calculating ω.   In fact, however, our new formalism is much more powerful than this; first, because we do not need to accept the semiclassical quantum state density formulas, and second because we can consider the possibility that rotational energy may also contribute to the critical coordinate energy ε_t.

Improved quantum state density formulas can be obtained by direct counting of vibrational–rotational states with a computer or by a variety of mathematical methods.   The contribution of one or more rotational degrees of freedom to ε_t can be taken into account also in several ways.   For example, a Laplace transform method for the case that r rotational degrees of freedom as well as s oscillators can contribute to ε_t gives in place of the semiclassical formulas for $r = 0$ a pair of formulas containing series expansions

$$g^*(\varepsilon) = \frac{q_r^*}{(kT)^{r/2} \prod_{i=1}^{s} h v_i} \sum_{\substack{i=0 \\ \text{even } i}}^{s} \frac{D_i(s)(\varepsilon + \varepsilon_z^*)^{s-i-1+r/2}}{\Gamma(s - i + r/2)}$$

$$N^+(\varepsilon) = \frac{q_r^+}{(kT)^{r/2} \prod_{i=1}^{s-1} h v_i^+} \sum_{\substack{i=0 \\ \text{even } i}}^{s-1} \frac{D_i(s - 1)(\varepsilon + \varepsilon_z^+)^{s-i-1+r/2}}{\Gamma(s - i + r/2)}$$

where

$$D_i^{(m)} \equiv \sum_{e=1}^{m} \Gamma(k_e + 1)^{-1}(h v_e/2)^{k_e} D_{k_e}$$

the sum and product being taken over all possible combinations of positive, even k_e for which

$$\sum_{e=1}^{m} k_e = i$$

and the D_{k_e} are coefficients related to Bernoulli numbers such that $D_0 = 1$, $D_2 = -1/3$, $D_4 = 7/15$, $D_6 = -31/21$, $D_8 = 127/15$.

EXERCISE 5-11 (a) For the case $r = 0$, show that the first term of the series expansion reproduces the semiclassical formulas. (b) This is a warning about series expansions. Evaluate the $i = 0$ and $i = 2$ terms in $g^*(\varepsilon)$ for $r = 0$, $s = 10$, $\varepsilon + \varepsilon_z^*$ = 20,000 cm^{-1}, all $\nu_i = 3 \times 10^{13}$ sec^{-1}. Note that these values for s, $\varepsilon + \varepsilon_z^*$, and ν_i are typical for a thermal unimolecular reaction calculation, and that extension to higher terms in the series will require a computer.

Other factors that can be included are corrections for anharmonicity of the A* vibrations, allowance for back reaction in the case of unimolecular isomerizations, improved treatment of the effect of rotation, and allowance for the possibility that there may be more than one path for a given reaction owing to isomeric forms of the critical configuration.

Regardless of which mathematical tools are used, the net result of the RRKM theory is a formula for k^1 containing a $k_a(\varepsilon)$ function based on a postulated model for the critical configuration and for the quantum state density for energized molecules. The agreement that can be obtained between RRKM theory and experimental results for k^1 is, therefore, contingent on an adventitious choice for these two experimentally inaccessible items of information. Therefore we must judge, to some extent, the success of applications of RRKM theory in the same spirit that we judge applications of ACT to bimolecular reactions: the models used must agree with our chemical intuition and be consistent with other information that we may have on the molecules under study. It is satisfying to find that the RRKM theory is successful using "reasonable" models for a wide variety of reactions. The fall-off curve for the reaction cyclopropane propylene shown in Figure 5-1 was computed using such a "reasonable" model; numerous other fall-off graphs among the reactions listed in Table 5-1 could be given. In fact, the only experimentally well-documented unimolecular reaction that has a k^1 in substantial disagreement with RRKM theory is $N_2O \rightarrow N_2 + O$, which has several complicating features, the most important of which is that the dissociation is not adiabatic; that is, it does not occur on a single potential energy surface (cf. legend to Figure 4-4).

Contrary to the situation with ACT of bimolecular reactions, acceptance of the random lifetime assumption for unimolecular reactions frees the RRKM theoretical framework from the "separability" objections to ACT mentioned in Section 4-7.

5-3 TESTS OF UNIMOLECULAR REACTION RATE THEORY

The RRKM theory is a coherent interpretation of the over-all energy transfer mechanism underlying unimolecular reactions. The k^1 function it yields does describe well the conversion rates of a variety of unimolecular reactions, and on this account we can be well satisfied with it. There are several individual components of the theory of energy transfer in unimolecular reactions that can be investigated separately by special experiments. We discuss several of these here, and one more in Chapter 10.

5-3a Isotope Effects

Vibrational frequencies in molecules depend in a predictable way upon the masses of their atoms. In diatomic molecules the relation is $v_0 = (C/\mu)^{\frac{1}{2}}/2\pi$ where C is the force constant and μ the effective mass of an oscillator; for polyatomic molecules the formulas are more complex. Moments of inertia also depend on atomic masses. The $k_a(\varepsilon)$ formula in RRKM theory therefore changes during isotopic substitution (e.g., of D for H), and the relative magnitudes of k^1 for a protonated and deuterated hydrocarbon reaction can be predicted easily. The few experiments available indicate satisfactory agreement. For example, the calculated ratio k_H^1/k_D^1 for the reaction $CH_3NC \rightarrow CH_3CN$ at 0.1 torr CH_3NC pressure is 0.35 ± 0.05 compared with theoretical estimates—depending on the model used for the critical configuration—ranging from 0.33 to 0.4.

5-3b Internal Activation

In a thermal unimolecular reaction, A* molecules are generated (according to our theory at least) in a Boltzmann vibrational distribution by energizing collisions. Boltzmann distributions have their merits, but others are possible and also useful. In unimolecular reactions, it is useful to generate A* molecules with energy ε much greater (i.e., many times kT greater, than the critical energy ε_0). The method of *chemical activation* is one way to achieve this. [*Photochemical activation* (Chapter 10) is another.] One synthesizes such highly energized A* by means such as reacting atoms (H) or diradicals (CH_2) with olefins. The resulting alkyl radicals or cyclopropanes are "*hot*" in the sense that the vibrational energy stored in them is sufficient to cause various unimolecular isomerizations or decompositions, provided that the vibrational energy of the hot molecules is not reduced below ε_0 by collisions. At low [M], therefore, one measures $k_a(\varepsilon)$ directly for $\varepsilon - \varepsilon_0 \gg kT$. The $k_a(\varepsilon)$ values found compare well with RRKM theory. The parameter s does turn out to be equal to the total number of vibrations of the hot molecules, and the random lifetime assumption as well as the computational procedures are vindicated. One significant finding is that hot methyl-cyclopropanes formed by the two reactions

$$CH_2 + \Delta \rightarrow CH_3 - \Delta^*$$

$$CH_2 + CH_2=CH—CH_3 \rightarrow CH_3—\Delta^*$$

decompose identically for the same value of $\varepsilon - \varepsilon_0$, even though the initial allocation of vibrational energy in $CH_3—\Delta^*$ is quite different for the two reactions. The excess energy moves around the $CH_3—\Delta^*$ molecules faster than they relieve themselves of it by decomposing.

5-3c   Energy Transfer to Diluents

In Table 5-2 we have collected some "per collision" relative energy transfer efficiencies for a number of M used as diluents for the thermal cyclopropane isomerization reaction. The essence of these results is that complex M's are more efficient in transferring energy. This can come about only if the internal degrees of freedom of M are able to couple with the internal degrees of freedom of A during an energizing or deenergizing encounter. Since the internal states of N-atomic M molecules that an A* might encounter are in a Boltzmann distribution with average vibrational energy $(3N - 6)kT$, which is much less than ε_0, strong coupling between the vibrations and rotations of M and A* implies high likelihood of deenergization, that is, validity of the strong collision assumption. The complementary reasoning holds also for energizing encounters $A + M \rightarrow A^* + M$. The correlation between efficiency and N of Table 5-2 is displayed qualitatively by other thermal unimolecular reactions.

Energy transfer to diluents can be studied in chemical activation and photo-chemical experiments as well. The same conclusion—high efficiency of energy transfer with increasing complexity of M—emerges. These results likewise inspire confidence in the validity of the strong collision assumption of thermal unimolecular reactions.

5-3d   Trajectories for Decomposition of Energized Molecules

Computer experiments can be done on unimolecular decompositions if the number of atoms per molecule is small, for example, 3 or 4. The procedure is essentially the same as for bimolecular $A + BC$ computer experiments. A model molecule is invented with bond force constants given by some analytic function that permits bonds to rupture if they have sufficient energy. Starting from any desired configuration (i.e., atom velocities and coordinates), we compute the evolution of the molecule by numerical integration of the classical equations of motion until reaction occurs, or until the computer decides that reaction will never occur.

Among the results of computer experiments on unimolecular decomposition are the following: the normal mode analysis used in vibrational spectroscopy and statistical mechanics proves to be quite useless for describing molecules energized to $\varepsilon > \varepsilon_0$. The amount of anharmonicity assumed in designing the decomposing molecule has a large effect on the results. All vibrations can contribute energy to the critical coordinate. Lifetime distributions are random unless some of the atomic masses or interatomic force constants are greatly different from the rest; this finding should not affect calculations for reactions of complex molecules, however, where $k_a(\varepsilon)$ becomes much larger than in the 3- and 4-atomic cases amenable to trajectory studies with digital computers.

Computer experiments on unimolecular decomposition also led to an amendment of the postulates of RRKM theory. Agreement between theoretical and computer decomposition rates was substantially improved if the value of the reaction coordinate of the critical configuration was such that the density of quantum states of the critical configuration is a minimum. There is, therefore, an unambiguous procedure for selecting the value of the reaction coordinate once the critical configuration for a given reaction has been identified.

The essence of the results of trajectory studies on unimolecular reactions is general corroboration of the assumptions and predictions of RRKM theory.

5-4 RÉSUMÉ OF UNIMOLECULAR REACTION RATE THEORY AND EXPERIMENT

The study of unimolecular gas reactions by painstaking experimentation on thermal reactions and nonthermal related reactions, imaginative construction of theory, and exploitation of digital computers has resulted in a pleasing scientific situation. In this area of chemical kinetics, decades of hard labor have produced genuine understanding of the observed facts. Future study will probably provide insights into the limits of validity of the assumptions of the existing theoretical framework and the appropriateness of computational methods attached to it rather than a more profound theory. On the experimental side, one hopes for specialized experiments testing the strong collision and the random lifetime assumptions as well as extension of fall-off experiments from large to smaller molecules and of computer experiments from small to larger molecules.

5-5 TERMOLECULAR GAS REACTIONS

We consider now *recombination reactions*, which have stoichiometric equations of the form

$$A + A \rightarrow A_2$$

where A is an atom, and their reverses, called *dissociation reactions*

$$A_2 \rightarrow A + A$$

We assume that the reactions are carried out at a high dilution in an inert gas M. In all cases the experimental rate law is found to be third order in the recombination direction and second order in the dissociation direction:

$$R_R = k_{R,M}(T)\,[A]^{2.0}[M]^{1.0}$$

$$R_D = k_{D,M}(T)\,[A_2]^{1.0}[M]^{1.0}$$

where we temporarily reinsert the temperature dependence of the rate constants to emphasize that theoretical understanding of these reactions is closely tied to understanding the experimentally determined temperature dependence.

Much of the experimental data on termolecular recombination reactions is obtained in flash photolysis experiments (Section 9-4), and virtually all of the data on bimolecular dissociation reactions come from shock tube experiments (Section 9-5). Chapter 9 should be consulted for a description of these experimental methods. Here we consider the results for two well-studied cases, the recombination reaction

$$I + I + M \xrightarrow{\;k_{R,M}\;} I_2 + M$$

for a large variety of M, called in this field of study the *third body* or *chaperon*, and the dissociation reaction

$$H_2 + Ar \xrightarrow{\;k_{D,Ar}\;} H + H + Ar$$

It is impossible to achieve and difficult even to approach experimental conditions in which only a single dissociation or recombination elementary reaction occurs. By varying the ratio of, for instance, I_2 and N_2 concentrations in a flash photolysis experiment on iodine atom recombination, the rate constants of the two parallel reactions

$$I + I + N_2 \xrightarrow{\;k_{R,N_2}\;} I_2 + N_2$$

and

$$I + I + I_2 \xrightarrow{\ k_{R,\, I_2}\ } I_2 + I_2$$

may be derived by assuming that the two proceed independently. We shall find later that considering dissociation and recombination reactions to be single elementary reactions is an oversimplification. Experimental k_R and k_D values are nonetheless derived on this basis.

The necessity of assuming that a third body participates in recombination reactions was recognized long ago on theoretical grounds. Consider the potential energy graph for the nonrotating H_2 molecule shown in Figure 5-2. If two H atoms approach one another head-on, their energy is necessarily *positive* with respect to the bound, nonrotating vibrational states of H_2.

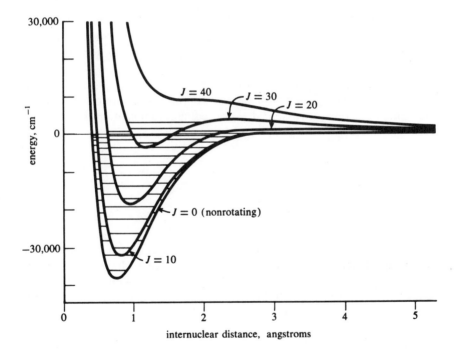

FIGURE 5-2 *Potential energy curves for the H_2 molecule in rotational states $J = 0$, 10, 20, 30, and 40. These rotational states have 15, 13, 9, and 4 vibrational states, respectively. The energies of the vibrational states are indicated by horizontal lines. Rotational states with $J = 35$, 36, and 37, have one vibrational state, and there are no bound states for $J > 37$. Potential energy curves for other diatomic molecules are similar to these except that the numbers of rotation and vibration states are larger. Potential energy curves to which rotational kinetic energy has been added are frequently called* effective *potential energy curves.*

The interatomic distance decreases to the point at which the potential energy $V(r)$ is equal to the initial relative kinetic energy of the approaching atoms; the direction of motion then reverses, and the two atoms part again with their original velocity vectors reversed. Formation of an H_2 molecule can only occur if a third atom or molecule intervenes when the two H atoms are close to one another and in some manner removes sufficient energy from the H-atom pair that their total energy is less than the dissociation energy. Here again the idea of energy transfer emerges as the critical factor in an elementary reaction. We write the corresponding mechanism, for example with an Ar atom as the third body, as a single elementary reaction

$$H + H + Ar \underset{k_{D,\,Ar}}{\overset{k_{R,\,Ar}}{\rightleftarrows}} H_2 + Ar$$

reserving for later consideration the observation that the H_2 molecule formed may be in any one of its many vibrational–rotational states. This one-reaction mechanism is called the *energy transfer mechanism* for dissociation and recombination reactions.

Another interpretation of the observed rate laws is possible, however, which is called the *complex formation mechanism*. In its simplest form three elementary reactions are assumed. For iodine recombination they would be

$$I + M \rightarrow IM^* \tag{1}$$

$$IM^* \rightarrow M + I \tag{2}$$

$$IM^* + I \rightarrow I_2 + M \tag{3}$$

The intermediate complex IM^* is unstable for the same reasons given in the preceding paragraph; however, it is now presumed to have a lifetime sufficiently long that IM^* survives until a later encounter with a second I atom leads to the formation of $I_2 + M$. In the dissociation direction, the complex formation mechanism assumes

$$I_2 + M \overset{k_{D,\,M}}{\longrightarrow} IM^* + I$$

followed by rapid bimolecular (mostly) or unimolecular decomposition of IM^*.

EXERCISE 5-12 The statistical mechanical formula for K_{eq} of $A_2 \rightleftarrows 2A$ in molecules cm^{-3} units is $K_{eq} = q_t^*(A)^2, q_e(A)^2, q_t^*(A_2)^{-1}, q_e(A_2)^{-1}, q_v(A_2)^{-1}, q_r(A_2)^{-1}$ $\exp(-D_0/RT)$ where the q's are partition functions and D_0 is the dissociation energy of A_2. For most dissociation–recombination problems the q_e are constants, $q_t^* = (2\pi mkT/h^2)^{\frac{3}{2}}$ and $q_r = 8\pi^2 IkTh^{-2}$. (a) Show that a graph of $\ln K_{eq}$ against T^{-1} has slope proportional to $D_0 + RT/2$ at low temperatures where $q_v = 1$ and proportional to $D_0 - RT/2$ at high temperatures where $q_v = kT/hc\omega_0$. (b) If $E_{A, R}$ for recombination is measured in a low temperature (300°K) experiment and $k_D/k_R = K_{eq}$ at all temperatures, how is $E_{A, D}$ measured at 3,000°K related to $E_{A, R}$? Compare Exercise 3-6.

In Figure 5-3 some flash photolysis $k_{R, M}$ data for I_2 recombination with various third bodies are shown. The wide variation of M efficiencies is striking, as is the fact that the E_A values are negative. In Figure 5-4 some experimental results for $k_{R, Ar}$ for H_2 recombination are shown. Once

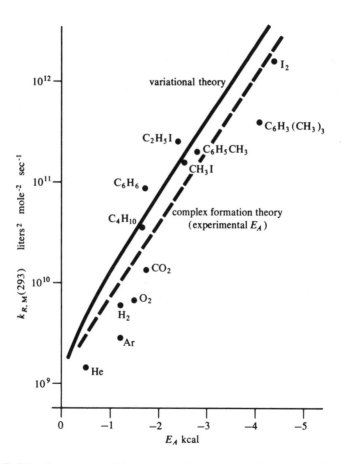

FIGURE 5-3   *Comparison of theoretical and experimental $k_{R, M}$ values for* $I + I + M \rightarrow I_2 + M$ *at 293°K. The theoretical lines represent the average behavior of the equations in the text.*

again it is found that k_R decreases with temperature, but here the measurements span the temperature range 300–5,000°K rather than 300–500°K, as was the case for the I_2 data of Figure 5-3.

EXCERISE 5-13   The $k_{R,Ar}$ data shown in Figure 5-4 are well represented by $k_R = 1.4 \times 10^{18} T^{-1}$ cm^6 mole^{-2} sec^{-1}.   (a) Convert this expression to a k_D expression in Arrhenius form.   At 1,000°K, $\log_{10} K_p$ for $H_2 \rightleftarrows 2H$ is -17.292 atm. Assume $E_B = D_0 = 4.476$ eV.   (b) Calculate $Z_{H_2, Ar}$ at 1,000°K assuming $\sigma_{H_2} = 1.3$ A, $\sigma_{Ar} = 1.8$ A.   (c) Calculate the exponent m in $(D_0/RT)^m$ and the steric factor p that would reconcile the activation-in-many-degrees-of-freedom rate constant expression (Section 4-3) with the experimental magnitude of k_D at 1,000°K.

The experimental findings on this class of reactions can be summarized as follows.   Most recombination reactions have rate constants that decrease with temperature approximately as T^{-1} or as T^{-2}.   The rates can depend strongly on the nature of the third body: in the case of I_2, $k_{R,M}$ values vary

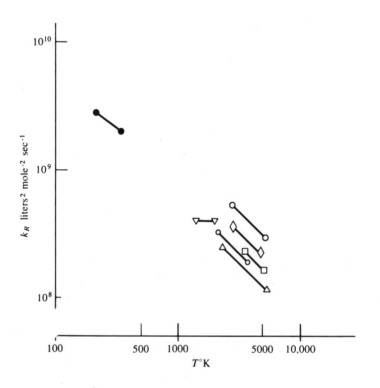

FIGURE 5-4   *Recombination rate constants for* $H + H + Ar \rightarrow H_2 + Ar$.   *The high temperature shock tube experiments measure* k_D; *the high temperature* k_R *values shown are calculated from* $k_R = k_D K_{eq}^{-1}$. *The solid lines are least-squares lines through shock tube data taken with five different shock tube methods.*

over 1,000-fold for different M. Large $k_{R,M}$ have greater temperature dependences. For simple, inert third bodies there is little variation among the recombination rate constants for all recombinations; they are almost all in the range 10^9–10^{10} liters^2 mole^{-2} sec^{-1}. For third bodies that can interact strongly with the recombining atoms, such as O_2 in O-atom recombination, k_R can be considerably enhanced. In shock tube measurements of dissociation rates, E_A is found to be many kilocalories less than D_0 except, apparently, for O_2. The A factors for k_D expressions in Arrhenius form are much greater—factors of 10 or more—than Z_{A_2M}. In shock tube experiments where excitation of lower vibrational states of A_2 can be observed as well as dissociation rates, it is found that vibrational excitation precedes dissociation.

A theory of termolecular gas reactions should provide a coherent framework for interpreting these findings on a molecular basis, as does the RRKM theory for unimolecular reactions. No such coherent framework exists. Instead, theoretical interpretations of termolecular reactions are only able to encompass parts of the body of available experimental results. Each of the many proposed theoretical efforts succeeds with part of the results, but ignores or is wrong about the rest. In the following sections we consider three different successful, but partial, theories of termolecular reactions.

5-6   VIBRATIONAL EXCITATION

One can imagine that an A_2 molecule could dissociate in three distinct ways: a molecule in a low v, J state could gain sufficient energy in a single, exceptionally energetic encounter to dissociate; alternatively, successive encounters could introduce sufficient rotational energy into the molecule that the centrifugal force overcomes the bond strength; or third, successive encounters could introduce sufficient vibrational energy into the molecule that the stretching force overcomes the bond strength. Reference to the potential energy curves (Figure 5-2) for rotating molecules shows why centrifugal dissociation is unlikely to be the principal route: much more energy is required to raise the molecule to an unbound J state than is required to raise a molecule in a low J state to an unbound v state. A decision on the relative importance of the first and third routes is possible also on theoretical grounds. For a model diatomic molecule with a finite number of bound vibrational states, for example a Morse oscillator, the probabilities for transitions between vibrational states caused by collisions (cf. Section 4-6) can be computed by essentially the same quantum mechanical methods used to calculate probabilities for transitions caused by light absorption or emission. When this is done, it is found that collision-induced transitions between vibrational states are far more probable than transitions from a bound vibrational state to the continuum of unbound states *except* for the few vibrational states that are just lower in energy than the dissociation limit.

Experimentally it was also found in shock tube experiments on O_2 dissociation that vibrational excitation precedes dissociation. We conclude, therefore, that the main route of dissociation is by vibrational excitation to high energy vibrational states, followed by collisional dissociation to atoms. In the reverse process of recombination, molecules are first formed in high energy vibrational states, then lose vibrational energy in later encounters to establish the Boltzmann distribution in vibrational states.

EXERCISE 5-14   An H_2 molecule in $v = 10$, $J = 10$ could dissociate by increasing J to 18 or v to 12.   Calculate the relative probability of these two routes at 3,000°K assuming that this is governed by the weighted Boltzmann factors.   Vibrational quantum states have statistical weight 1, whereas rotational quantum states have statistical weight $2J + 1$.   Energies $E_{v, J}$ in cm^{-1} of these states are $E_{10, 10} = 33,966$, $E_{10, 18} = 37,000$, and $E_{12, 10} = 36,106$.

The simplest dissociation rate theories that we can regard as steps in the right direction analyze the rate of the successive excitations to higher energy vibrational states, ignoring the obvious complications that consideration of rotation and distortion of the $V(r)$ curve during the exciting collisions would introduce. These are called *vibrational excitation* theories, or, from the appearance of the vibrational energy levels shown in Figure 5-2, *ladder-climbing* theories, or from the mathematics employed, *stochastic* theories. The dissociation mechanism for an A_2 molecule with n vibrational states then becomes

$$A_2 \, (v = 0) \underset{k_{10}}{\overset{k_{01}}{\rightleftarrows}} A_2 \, (v = 1)$$

$$A_2 \, (v = 1) \underset{k_{21}}{\overset{k_{12}}{\rightleftarrows}} A_2 \, (v = 2)$$

$$A_2 \, (v = n - 1) \underset{k_{n, n-1}}{\overset{k_{n-1, n}}{\rightleftarrows}} A_2 \, (v = n)$$

$$A_2 \, (v = n) \underset{k_{D, n}}{\overset{k_{n, D}}{\rightleftarrows}} 2A$$

if it is assumed that dissociation occurs only from the nth vibrational state. The first order rate constants indicated in the mechanism represent second order rate constants multiplied by [M], since an M is understood to occur on both sides of each equation. It can be assumed that k_{01} is known from an experiment on vibrational excitation. The ingredients of the theory are twofold: calculation of the other k values and then calculation of $d[A]/dt$.

Unfortunately, advanced mathematics and voluminous algebra are required to develop vibrational excitation theories; so we must be content merely to note two of the important results. First, it proves impossible to obtain large enough dissociation rates if the k values are derived from the theory of harmonic oscillators. It is necessary to assume that the anharmonicity is important in increasing the k values for transitions among the high energy vibrational levels. Experimental determinations of k values for these levels thereby become of prime importance for dissociation theory.

A second major result is obtained by numerical integration of the complete set of n rate equations. It is found that the lower vibrational states are populated according to a Boltzmann distribution during the first part of the dissociation process—before $k_{D,n}$ $[A]^2$ is important—while the highest vibrational states are not. The experimental finding that E_A is less than D_0 can thus be interpreted as implying that molecules excited to vibrational states near D_0 are essentially atomized as far as the kinetics of the early dissociation process are concerned.

Vibrational excitation theory is one reasonable way to think about the dissociation process; as a scientific theory, however, it is quite inadequate because of its large number of inaccessible parameters (the k values) and its neglect of rotation and of the distortion of $V(r)$ during encounters. Since this last omission precludes consideration of the dependence of $k_{R,M}$ on the nature of M, a major part of the experimental findings is left uninterpreted.

5-7 "TERMOLECULAR" AND "COMPLEX FORMATION" RECOMBINATION THEORY

We divide the recombination into two paths

$$A + A \text{ "$\rightleftarrows$" } A_2^* \tag{1}$$

$$A_2^* + M \rightarrow A_2 + M \tag{2}$$

and

$$A + M \rightleftarrows AM \tag{3}$$

$$AM + A \rightarrow A_2 + M \tag{4}$$

which can be called the *termolecular* and *complex formation* pathways, respectively. A_2^* represents an unbound, temporary pairing of A atoms that must lose energy to M if recombination is to occur. As we have seen in the previous section, the A_2 molecules contain a large amount of vibrational energy, but we now focus our attention elsewhere. Considering the two routes as parallel reactions, the rate law

$$R = R_T + R_{CF} = (k_T + k_{CF}) [M][A]^2$$

is assumed to describe the recombination kinetics.

A steady state assumption on $[A_2^*]$ yields for the termolecular path

$$R_T = \frac{k_1 k_2 \, [M][A]^2}{k_{-1} + k_2 \, [M]}$$

Since redissociation (-1) is far more likely than stabilization (2) at all experimentally accessible pressures (in contrast to the situation in unimolecular dissociation of polyatomic molecules), the second term in the denominator can be neglected. In place of k_1/k_{-1} we write K_1, and we factor k_2 into a probability P and an encounter frequency $Z_{A_2^*M}$. (In this section we consistently give Z a subscript and assume that it is in liter mole^{-1} sec^{-1}.) Then

$$k_T = P Z_{A_2^*M} \, K_1$$

The simplest assumption to make about P is a strong collision assumption, $P = 1$. $P < 1$ may also be reasonable, for example, if there are unbound $V(r)$ to consider. The encounter number $Z_{A_2^*M}$ can be calculated with estimated collision diameters (cf. Section 4-2). K_1 is, in this case, *not* an equilibrium constant in the usual sense since A_2^* is not a bound molecule. It is, however, still given by the ratio $[A_2^*]/[A]^2$, and can be calculated as follows. Let $Z_{AA}(\varepsilon, b) \, [A]^2 \, d\varepsilon \, db$ be the number of A–A encounters per liter per second with relative energy in the range ε to $\varepsilon + d\varepsilon$ and impact parameter in the range b to $b + db$, and let $\tau(\varepsilon, b)$, which we shall call the *lifetime* of A_2^* for this range of ε and b, be the length of time that the A–A distance r in these encounters will be small enough so recombination can occur. Then

$$[A_2^*] = \iint Z_{AA}(\varepsilon, b) \, [A]^2 \tau(\varepsilon, b) \, d\varepsilon \, db$$

$$K_1 = \iint Z_{AA}(\varepsilon, b) \tau(\varepsilon, b) \, d\varepsilon \, db$$

The time $\tau(\varepsilon, b)$ during which r is in the range r_1 to r_2 is given by

$$\tau(\varepsilon, b) = (\mu_{AA}/2)^{\frac{1}{2}} \int_{r_1}^{r_2} [\varepsilon \, (1 - b^2/r^2) - V(r)]^{-\frac{1}{2}} \, dr$$

For any $V(r)$ function, the value of K_1 corresponding to any assumed limit on r may be calculated by doing these integrals on a computer. Once this is done, k_T can be computed from $k_T = P Z_{A_2^*M} \, K_1$ and compared with experiment.

For iodine atom recombination, taking $P = 1$ gives $k_T(300) \approx 10^9$ liters^2 moles^{-2} sec^{-1}. Therefore we can accept $k_{R,M} \approx k_T$ for inert gas atoms. Experimentally, k_R decreases with temperature; k_T calculated as indicated above, however, increases. Moreover, there is no way to account for the larger $k_{R,M}$ values in the termolecular theory since P already has its maximum value of unity. We conclude that the data in Figure 5-3 cannot be accounted for by the termolecular pathway alone.

The rate equations for the complex formation pathway can be treated like those for the termolecular pathway, except that in

$$k_{CF} = PZ_{AMA} K_3$$

the quantity K_3 can be regarded as a true equilibrium constant since AM is bound, at least weakly. Two procedures have been used to calculate K_3. One of them accepts the experimental E_A to obtain the binding energy $D(A-M)$ and uses partition functions to obtain K_3, and hence k_{CF}, for the various M of Figure 5-3. The absolute values of k_{CF} calculated for $P = 1$ are in good agreement with experimental $k_{R,M}$ values, as shown in Figure 5-3. An alternative procedure is to compute K_3 from van der Waals constants and other data concerning the A-M interaction. In this case both the value of $k_{R,M}$ and the temperature dependence can be compared with experiment. The agreement is good (cf. Exercise 5-15).

We conclude that atom recombination by atomic third bodies can proceed with contributions from both the termolecular and the complex formation paths, whereas recombination by polyatomic third bodies proceeds predominantly by the complex formation path. Considering recombination in terms of $k_R = PZ_{A_2^*M} K_1 + PZ_{AMA} K_3$ provides a suitable rationalization of the magnitude of $k_{R,M}$, the relative magnitudes of $k_{R,M}$ for different M, and their temperature dependence. This theoretical approach lacks only the possibility of rationalizing for the reverse reaction of dissociation the observed interaction of vibrational excitation and dissociation into atoms.

EXERCISE 5-15 An approximate formula for K_3 in the complex formation theory is $K_3 = \pi^{\frac{1}{2}} \sigma^3 (\varepsilon/kT)^{\frac{3}{2}} (8/3 + (32/45) (\varepsilon/kT))$ where σ and ε are Lennard–Jones 6–12 constants. Select one of the M from Figure 5-3 and use ε and σ values to calculate a $k_{R,M}$, E_A point for Figure 5-4. Obtain ε and σ from a textbook on the kinetic theory of gases (cf. Supplementary References to Chapter 4) and assume that ε and σ for I are the same as they are for Xe. *Hint:* Remember that $E_A \equiv -R[d \ln k_R/d(1/T)]$.

5-8 VARIATIONAL THEORY OF TERMOLECULAR REACTION RATES

We have seen that including the effects of A-M interaction in the complex formation theory allows satisfactory rationalization of the M dependence of the experimental $k_{R,M}$ data for iodine-atom recombination. Now we consider a theoretical framework which simultaneously includes *all* atom–atom and atom–molecule interactions in a recombination reaction. To achieve this, a trial surface in the three-particle (A, A, M) phase space, chosen to divide the phase space into bound and unbound regions, is systematically varied until the lowest recombination rate, corresponding to a flux of phase points across the trial surface, is obtained. The name *variational theory* is derived from this procedure. Since the phase points can pass through the trial surface from the bound to the unbound region and out again, this phase point flux is an upper bound for the recombination rate. In a separate study the fraction of multiple crossings of the trial surface is found by trajectory calculations; it turns out to be a simple function of the difference between the rotational and vibrational energy at a given location in the phase space. Correcting the phase point flux for the actual fraction of crossings that result in recombination yields an absolute calculation of $k_{R, M}$.

The algebra of the variational theory is again rather extensive; so we must again be content to consider the ideas involved without the corresponding equations. The phase space for the three-atom system A, A, M is $6N = 18$-dimensional including the c.m. motion, or 12-dimensional ignoring it. The classical Hamiltonian function for the system, which contains the $V(r)$ functions for the A–A and A–M interactions, determines the evolution of phase points. By suitable coordinate transformations the phase space is expressed in terms of r_3, the M to A–A midpoint distance, which is associated with a weakly attractive $V(r_3)$ function, and the A–A distance r_{12}, which is associated with a Morse $V(r_{12})$ function reproducing the known properties of the A_2 molecule; the remaining four coordinates are angles. The trial surface is composed of two parts, analogous to the separation of k_R into k_T and k_{CF}. It is defined by two equations combining particular values of r_3, r_{12}, and $\mu_{A_2} \dot{r}_{12}^2/2$, which are assumed to distinguish bound A_2 from unbound $A + A$; the value of r_3 is chosen as the parameter to be varied. A computer integration is needed to calculate the flux and minimize it with respect to the value of r_3 defining the trial surface.

Combining the computer results for the flux and for the fraction of recombining crossings then provides $k_{R, M}$. For inert gas $k_{R, M}$ calculations the necessary constants for the $V(r_3)$ function can be estimated easily, and $k_{R, M}$ can be calculated directly. The results are in good agreement with experiment. For polyatomic M's, the theory can be compared indirectly with experiment, again with good results.

EXERCISE 5-16 The variational theory yields for iodine–atom recombination the expression $k_{R, M} = 2.0 \times 10^8 \, \sigma_M{}^2 \, (\exp(\varepsilon_M/RT) - 0.6)$ liters^2 moles^{-2} sec^{-1}, where σ_M and ε_M are parameters of the $V(r_3)$ function in angstrom and kcal units, respectively. (a) What activation energy is implied by this functional form? (b) Compute σ_M at 300°K for M = Ar, O_2, CO_2, and I_2 from the data in Figure 5-3. *Hint:* Construct a graph of ε versus E_A from your answer to part (a). For comparison, the values of σ_M from viscosity measurements are $\sigma_{Ar} = 3.6$, $\sigma_{O_2} = 3.8$, $\sigma_{CO_2} = 4.0$, and $\sigma_{I_2} = 4.2$ Å.

The variational theory has, in addition to its success in rationalizing the iodine data, the advantage that no assumptions need be made about encounter cross sections, which are necessary to compute Z values in other theories, since $k_{R, M}$ here depends only on the interaction potentials.   Also, it considers all paths leading to recombination.   Its only shortcomings are lack of consideration of the structure of M, and, in the version given here, lack of consideration of the vibrational transitions in A_2.

5-9   ENERGY TRANSFER IN TERMOLECULAR GAS REACTIONS

Energy transfer enters into theories of termolecular reactions in three ways, each offering different questions to theoreticians and experimentalists.

First there is the transfer of energy between relative translational energy of two A atoms and relative translational energy of a bound, but vibrationally excited, A_2 molecule and an atomic third body M.   This energy transfer is the cornerstone of that part of termolecular reaction theory that we found to be adequate for explaining the magnitudes, but not the temperature dependences, of $k_{R, M}$ for recombination by atomic M.   Given the interaction potentials $V(r)$, the theory of this kind of energy transfer is a problem in collision mechanics that we can regard as solved.   Information about the pertinent interatomic interactions, however, is sparse.   For very small atoms—helium and hydrogen—the $V(r)$ functions have been calculated accurately from quantum mechanics.   For other atoms, the only available information is in the form of $V(r)$ functions derived from virial coefficient or transport property measurements.   Since the encounters that determine the nonideality or the transport properties of gases are more energetic than the encounters that dominate in termolecular reactions, these data are of doubtful utility for termolecular reaction rate calculations.

The second kind of energy transfer is between recombining atom and diatomic or polyatomic third bodies.   We can subdivide this into transfer between the translational energies of these third bodies and the recombining atoms and transfer between the internal energy of the third body and the recombining atoms.   The latter resembles energy transfer within energized polyatomic molecules undergoing unimolecular reaction; so theoretical and experimental evidence about intramolecular energy transfer in energized molecules sheds light on the mechanics of the complex formation pathway of termolecular recombination.   Transfer of translational energy from recombining atoms to translational energy of a diatomic or polyatomic third body resembles the transfer to an atomic third body except that the interaction potentials are different.

The third kind of energy transfer refers to changes of the vibration and rotation states of energized diatomic molecules resulting from encounters other than ones which actually accompany bond formation in recombination or bond rupture in dissociation. In order to account for the temperature dependence of the rate constants or the observed coupling of vibrational excitation and dissociation, this energy transfer must be considered to play an important role. There are few theories and experiments that bear on it. From the vibrational excitation theory we found that experimental information about energy transfer within the low vibration and rotation states is inadequate to describe energy transfer within highly energized states. Experimental confirmation of this finding comes from fluorescence quenching experiments on diatomic molecules in excited electronic states (Section 10-6).

The study of termolecular recombination reactions, and their reverse reactions of dissociation, is a diverse and fruitful field of scientific inquiry where we are able to touch, but not yet grasp, explanations of elementary reaction rates in terms of atomic and molecular properties. Theoretical and experimental research here has much the same flavor as research on the bimolecular atom plus diatomic molecule exchange reactions discussed in Chapter 4.

SUPPLEMENTARY REFERENCES

Unimolecular and termolecular rate constant data have been tabulated by

S. W. Benson and H. E. O'Neal, *Kinetic Data on Gas Phase Unimolecular Reactions*, U.S. National Bureau of Standards Report NSRDS-NBS-21, 1970.

K. Schofield, *Planetary and Space Science* **15**, 643 (1967).

J. Troe and H. Gg. Wagner, "Unimolekulare Reaktionen in thermischen Systemen," in *Berichte der Bunsengesellschaft* **71**, 937 (1967).

The data on iodine recombination are discussed in several papers in

Inelastic Collisions of Atoms and Simple Molecules; Discussions of the Faraday Society **33**, 1962.

Unimolecular and termolecular reaction theory is discussed by

D. L. Bunker, *Theory of Elementary Gas Reaction Rates* (Pergamon Press, Oxford, 1966), Chapters 3 and 4.

H. S. Johnston, *Gas Phase Reaction Rate Theory* (Ronald Press Co., New York, 1966), Chapters 14 and 15.

E. E. Nikitin, *Theory of Thermally Induced Gas Phase Reactions* (Indiana University Press, Bloomington, Indiana, 1966).

N. B. Slater, *Theory of Unimolecular Reactions* (Cornell University Press, Ithaca, N.Y., 1959).

The statistical mechanical formulas used in this chapter are derived in

L. S. Kassel, *The Kinetics of Homogeneous Gas Reactions* (Chemical Catalog Company, New York, 1932), Chapter II.

A description of the role of rotation in unimolecular reactions is given by

R. A. Marcus, "Unimolecular Reaction Rate Theory," in *Chemische Elementarprozesse*, H. Hartmann, Ed. (Springer Verlag, Berlin, 1968).

Chapter 6

REACTIONS IN SOLUTION

IN THE previous two chapters we found that mechanisms of gas reactions have elementary steps that are isolated encounters between individual molecules, and that the theory of gas reactions is the theory of such isolated encounters. Chemical reactions are mostly carried out in liquid solutions, however, in which the notion of isolated encounters is meaningless. Reactant molecules interact continuously with solvent molecules, in the transition state and otherwise, and this fact causes changes in the observed phenomena and in the nature of theoretical interpretations as well.

A general characteristic of solution kinetics is that the observed conversion rates are identified much more directly with single transition states than in gas kinetics. This facilitates understanding the rate limiting steps, which is the main interest of the organic chemist or the chemical engineer, but obscures the rest of the mechanism, which the physical chemist would like to know. The relative ease of identifying conversion rates with transition states also facilitates the application of the activated complex theory, the premises of which may be more nearly valid for solution reactions than for gas reactions. Specialized methods, discussed in Chapter 9, are needed to study elementary steps of solution reactions that are too fast to influence conversion rates.

Solvent effects are classified as *physical* when the solvent molecules intrude into a mechanism by permitting reactant molecules to have elementary reactions different from those which they have in the gas phase, and as *chemical* when the solvent molecules themselves appear in the mechanism. By far the most important physical effect of solvation is ionization; ions appear rarely in mechanisms of gas reactions but frequently in mechanisms of solution reactions. A second important physical effect is that rapid energy transfer in the abundant collisions between reactant and solvent molecules maintains thermal, in particular vibrational, equilibrium at all times. The role of vibrational energy transfer so prominent in the theory of unimolecular and termolecular reactions in gases is therefore entirely absent from the theory of solution reactions. A third physical effect, discussed in Section 6-2, is that interactions between reacting ions depend on the electrical environment of the ions and therefore can be influenced by the dielectric character of the solvent and by the presence of other ions in the solution. Chemical effects are of two kinds: the solvent may have a catalytic role in the mechanism and be regenerated as fast as it is consumed; or it may appear in the stoichiometric equation as a reactant or product and be permanently consumed or generated.

There are a few reactions that have the same mechanism in a variety of solvents as they have in the gas phase. For these reactions, almost the same rates and Arrhenius parameters are measured in solutions as in the gas phase, thus appearing to imply that encounter frequencies in solutions are about the same as in gases. Reconciling this fact with the great difference in density between the two states of matter is left as an exercise for the reader (cf. Exercise 6-16). We see later that the equivalence cannot hold for elementary reactions that occur at every encounter (cf. Exercise 6-8).

Research on solution reaction mechanisms usually proceeds far beyond rate measurements and assignment of a set of elementary reactions. Of particular interest are the stereochemistry of the transition states involved, the nature of the bonding changes that occur during reaction, and the variations of reactivity among related reactions. In order to provide a proper presentation of the knowledge that has been gained concerning the details of transition states and relative reactivities in, for example, substitution reactions of one ligand for another at a metal ion in aqueous solution, we would have to develop a large amount of descriptive chemistry and bonding theory. Since this would greatly enlarge the scope of this book, we have to forego discussing these important aspects of chemical kinetics. They are thoroughly treated in organic and inorganic chemistry courses and in the references cited at the end of the chapter. Here we confine ourselves to introducing four major topics of solution kinetics as a means of placing the subject of solution reactions into context with the rest of chemical kinetics.

6-1 THEORETICAL INTERPRETATION OF RATE CONSTANTS FOR ELEMENTARY REACTIONS IN SOLUTION

Elementary reactions in liquid solutions can be unimolecular, bimolecular, or termolecular. Just as in the case of gas reactions, each of the three types of elementary reaction has a different form of theoretical interpretation. The

interpretations are significantly different from the corresponding interpretations of gas phase elementary reactions.

6-1a Unimolecular Reactions

In the high pressure limit of gas phase unimolecular reactions, the first order
rate constant was found (cf. Section 5-2) to be the product of a rate constant
for unimolecular decomposition and an equilibrium constant describing the
equilibrium between reactant molecules and energized reactant molecules

$$k_{\infty} = K_{eq} k_{uni}$$

The high density of liquids ensures that the energization–deenergization
equilibrium condition *always* holds for reactions in solution.   Therefore one
can begin the theory of unimolecular reactions in solution by assuming that
calculation of k_{∞} is the primary goal.   Based on the formalism of RRKM
theory (Section 5-2c), we would anticipate that k_{∞} for a unimolecular reaction
in solution can be calculated theoretically using an assumed density of states
formula and the experimental activation energy.   This is indeed a reasonable
procedure.   The main objection that one would make to it is that the density
of states function could be significantly affected in an unknown way by
interactions between energized A* molecules and the surrounding solvent M
molecules.

The thermodynamic formulation of activated complex theory (Section 6-1c)
is also used in discussing rate constants for unimolecular reactions in solution.

Unimolecular decompositions in solution have a mechanistic complication
called the *cage effect* that is absent from the corresponding gas reactions.   The
products of a unimolecular decomposition in solution are initially formed
right together and stay close to one another within a cage of solvent molecules
until diffusion brings solvent molecules between them.   Before the solvent
molecules intervene and the product molecules or radicals escape from the
cage in which they were formed, they have a good chance of reacting with one
another and thus avoiding reaction with other species present in the solution.
For example, methyl radicals formed in solution by thermal unimolecular
decomposition of azomethane

$$CH_3N_2CH_3 \rightarrow N_2 + 2CH_3$$

mostly recombine to form ethane

$$CH_3 + CH_3 \rightarrow C_2H_6$$

within the solvent cage in which they were formed and thereby avoid other reactions that they might have undergone had they diffused away from one another.

6-1b Termolecular Reactions

In atom recombination reactions in the gas phase we reasoned (Section 5-5) that a bimolecular reaction

$$A + A \rightarrow A_2^*$$

could not lead to formation of diatomic A_2 molecules unless a third body M were to encounter A_2^* within a single vibration period to prevent redissociation into atoms. In solution the solvent molecules are always available for energy transfer, and this reasoning no longer holds. The rate constant for atom (or radical) recombination in solution is determined by the rate of diffusion through the solvent, a subject to which we return in Section 6-5.

Termolecular reactions in solution are thus expected to be rare, if they exist at all. One can usually regard third order processes in solution as caused by a set of successive bimolecular reactions. They might involve, for example, equilibria preceding product formation, such as

$$A + B \rightleftharpoons AB$$

$$A + C \rightleftharpoons AC$$

$$B + C \rightleftharpoons BC$$

$$\left. \begin{array}{l} C + AB \rightarrow \\ B + AC \rightarrow \\ A + BC \rightarrow \end{array} \right\} \text{products}$$

A true termolecular reaction in solution would be the limiting case in which the equilibrium constants for all three equilibria were very small, which means that the intermediates AB, AC, and BC would have very short lifetimes. The experimental third order rate constant in $R = k[A][B][C]$ would represent the largest product of equilibrium constant and bimolecular rate constant from the successive reactions. Since all reactants and intermediates are solvated species, the energy transfer processes associated with the theory of termolecular reactions in the gas phase do not play an important role in the theory of third order solution reactions. Instead, a theoretical interpretation would consider the lifetimes of the intermediate species and the rates of their further reactions from the points of view of unimolecular and bimolecular rate theory.

6-1c   Bimolecular Reactions—Thermodynamic Formulation of Activated Complex Theory

By far the most important elementary reactions in solution are the bimolecular ones, and the theory of solution reactions is therefore mainly the theory of bimolecular solution reactions. Some examples of bimolecular solution reactions are given in Table 6-1. Now we wish to introduce the theoretical concepts generally used to interpret the Arrhenius parameters of reactions such as these.

The concept of cross section, through which we developed the theory of bimolecular gas reactions, is unfortunately not of any use to us here; for in bimolecular solution reactions each bimolecular encounter is a complex sequence of collisions rather than a single scattering event. The presence of solvent molecules, which destroys the usefulness of the reactive scattering concept, also prevents us from utilizing the activated complex theory (ACT) of bimolecular gas reactions (Section 4-7). This occurs because the partition functions per unit volume in the ACT rate constant formula can be computed only for isolated molecules, not for solvated molecules. We now develop an alternative ACT formalism, based on thermodynamic considerations, that will be applicable to solution reactions.

For bimolecular gas reactions of the form

$$A + B \rightarrow D + E$$

the ACT rate constant formula was found to be

$$k(T) = \frac{kT}{h} \frac{q_C^{*\dagger}}{q_A^* q_B^*} e^{-E'_b/kT}$$

We recall that the factor kT/h was obtained by taking the product of the average translational velocity in one direction on the reaction coordinate and the number of activated complexes per unit length of the reaction coordinate, with the partition function for the degree of freedom representing the reaction coordinate having been factored out of the total partition function per unit volume q_C^*. We now regard kT/h as the rate of product formation for unit-activated complex concentration. Since $k(T)$ itself is the rate of product formation for unit reactant concentrations, the factors multiplying kT/h convert the reactant concentration product $^1N_A \, ^1N_B$ to the activated complex concentration ^1N_C; that is, they comprise an equilibrium constant. In applying ACT to a gas reaction, we postulate the properties of an activated complex and compute this equilibrium constant from partition functions.   In applying ACT to a solution reaction, we must use a different procedure since the partition functions are not known.   The equilibrium constant for $A + B \rightleftharpoons C$ is related to a free energy change $\Delta G^{0\dagger}$ by the thermodynamic equation

$$K_{eq}^{\dagger} = \exp(-\Delta G^{0\dagger}/RT)$$

$$= \exp(\Delta S^{\dagger}/R)\exp(-\Delta H^{\dagger}/RT)$$

where $\Delta G^{0\dagger}$ is called the *free energy of activation*, $\Delta S^{\dagger}$ is called the *entropy of activation*, and $\Delta H^{\dagger}$ is called the *enthalpy of activation*.   The rate constant equation is then

$$\boxed{k(T) = \frac{kT}{h}\, e^{\Delta S^{\dagger}/R} e^{-\Delta H^{\dagger}/RT}}$$

If it were possible to calculate $\Delta S^{\dagger}$ and $\Delta H^{\dagger}$ values from molecular properties, then $k(T)$ could be computed.   Unfortunately, such a computation is an uncertain one for solution reactions on account of the uncertain effects of the solvent on the thermodynamic properties of a postulated activated complex.   For this reason ACT is mostly used for interpreting rate constants for solution reactions rather than for attempting *a priori* calculations of rate constants.   The first step in such an interpretation would be to compute $\Delta S^{\dagger}$ and $\Delta H^{\dagger}$ from the Arrhenius parameters of the reaction; the second would be to compare these values with those found for related reactions.   *A priori* rate constant calculations for solution reactions are only useful for special cases, such as the case of a reaction that involves only the transfer of an electron from one ion to another.

EXERCISE 6-1   (a) What are the thermodynamic standard states appropriate to the quantities $\Delta S^{\dagger}$ and $\Delta H^{\dagger}$?   (b) Derive formulas for computing $\Delta S^{\dagger}$ and $\Delta H^{\dagger}$ (in cal deg^{-1} mole^{-1} and kcal mole^{-1}) from the Arrhenius parameters of a rate constant expressed in liters mole^{-1} sec^{-1}.

EXERCISE 6-2   Calculate $\Delta S^{\dagger}$ and $\Delta H^{\dagger}$ values listed for one of the reactions in Table 6-1.   Explain the sign of your value for $\Delta S^{\dagger}$ in terms of a plausible activated complex structure.

Enthalpy of activation values are useful for thermochemical interpretations of the structure of transition states.   In essence, the enthalpy of activation is related to energy changes along the reaction coordinate as shown in Figure 3-1, and can be interpreted in terms of bond strengths as described in Section 3-5.

TABLE 6-1 RATE CONSTANTS FOR SOME SECOND ORDER SOLUTION REACTIONS

Reaction	k	A	E_A
$e^-_{aq} + e^-_{aq} \rightarrow H_2 + 2OH^-$	1×10^{10}		
$e^-_{aq} + H^+_{aq} \rightarrow H$	2.1×10^{10}		
$e^-_{aq} + OH \rightarrow OH^-$	3×10^{10}		
$e^-_{aq} + Cd^{2+} \rightarrow Cd^+$	5×10^{10}		
$e^-_{aq} + NH^+_4 \rightarrow H + NH_3$	1×10^6		
$e^-_{aq} + CO_2 \rightarrow CO^-_2$	8×10^9		
$H^+_{aq} + HS^- \rightarrow H_2S$	7.5×10^{10}		
$H^+_{aq} + NH_3 \rightarrow NH^+_4$	4.3×10^{10}		
$H^+_{aq} + N(CH_3)_3 \rightarrow HN(CH_3)^+_3$	2.5×10^{10}		
$H^+_{aq} + CuOH^+ \rightarrow Cu^{2+}$	1×10^{10}		
$H^+_{aq} + CH_3OH \rightarrow CH_3OH^+_2$	1×10^8		
$H^+_{aq} + C_2H_5OH \rightarrow C_2H_5OH^+_2$	3×10^6		
$H_3O^+ + H_2O \rightarrow H_2O + H_3O^+$	1×10^{10}		
$H_3O^+ + H_2O_2 \rightarrow H_2O + H_3O^+_2$	2×10^7	1.4×10^9	2.6
$OH^- + HCO_3^- \rightarrow CO_3^{2-} + H_2O$	6×10^9		
$OH^- + HPO_4^{2-} \rightarrow PO_4^{3-} + H_2O$	2×10^9		
$OH^- + NH^+_4 \rightarrow NH_3 + H_2O$	3.4×10^{10}		
$OH^- + EDTA^{3-} \rightarrow EDTA^{4-} + H_2O$	3.8×10^7		
$OH^- + H_2O \rightarrow H_2O + OH^-$	5×10^9	1.2×10^{13}	4.8
$OH^- + CH_3OH \rightarrow CH_3O^- + H_2O$	3×10^6		
$OH^- + C_2H_5OH \rightarrow C_2H_5O^- + H_2O$	3×10^6		

Reaction	k	A	E_A
$OH^- + p\text{-}C_6H_4(COOC_2H_5)_2 \rightarrow p\text{-}C_2H_5OOCC_6H_4COO^- + C_2H_5OH$	5.4×10^{-2}	2×10^8	13
$OH^- + \text{cyclopentanone} \rightarrow C_4H_9COO^-$	4.7×10^{-1}	1.3×10^6	11.2
$Co(NH_3)_5N_3^{2+} + Co(CN)_5^{3-} \rightarrow Co(NH_3)_5^{2+} + Co(CN)_5N_3^{4-}$	3×10^5		
$Co(NH_3)_5NCS^{2+} + Co(CN)_5^{3-} \rightarrow Co(NH_3)_5^{2+} + Co(CN)_5NCS^{4-}$ (308°K)	2.8×10^1		
$CH_3Br + Cl^- \rightarrow CH_3Cl + Br^-$ (in acetone)	5.9×10^{-3}	2×10^9	15.7
$CH_3Br + Br^{*-} \rightarrow CH_3Br^* + Br^-$ (acetone, 238°K)	6.3×10^{-4}	5×10^{10}	15.8
$trans\text{-}ICH{=}CHI + I^- \rightarrow C_2H_2 + I_3^-$ (CH$_3$OH, 352°K)	1.8×10^{-6}	4×10^{12}	29.4
$(CH_3)_3CCl + C_6H_5S^- \rightarrow (CH_3)_2C{=}CH_2 + C_6H_5SH + Cl^-$ (C$_2$H$_5$OH, 318°K)	4.1×10^{-5}	4×10^{12}	25
$CH_2{=}CHCH_2Cl + S_2O_3^{2-} \rightarrow CH_2{=}CHCH_2S_2O_3^- + Cl^-$ (50% C$_2$H$_5$OH)	2.5×10^{-3}	1.2×10^8	14.5
$C_6H_5N(CH_3)_3^+ + CH_3O^- \rightarrow C_6H_5N(CH_3)_2 + (CH_3)_2O$ (CH$_3$OH, 343°K)	9.7×10^{-5}	7×10^{16}	32.7
$C_6H_5CHO + HCN \rightarrow C_6H_5CH(OH)CN$ (C$_2$H$_5$OH, 293°K)	2.0×10^{-8}		
$CH_3I + C_6H_5N(CH_3)_2 \rightarrow C_6H_5N(CH_3)_3^+ + I^-$ (CH$_3$OH)	5.6×10^{-5}	7.9×10^6	15.2
$CH_3I + C_6H_5N(C_2H_5)_2 \rightarrow C_6H_5N(C_2H_5)_2CH_3 + I^-$ (CH$_3$OH, 338°K)	1.2×10^{-4}	5×10^7	18
fumarase + fumarate $\rightarrow$ enzyme-substrate complex	$>1 \times 10^9$		
catalase + $H_2O_2 \rightarrow$ enzyme-substrate complex	5×10^6		
hemoglobin$\cdot 3O_2 + O_2 \rightarrow$ hemoglobin$\cdot 4O_2$	2×10^7		

Notes to Table 6-1: All rate constants and A factors are in liters mole^{-1} sec^{-1} units. Unless otherwise indicated, the reactions were studied in water solution at 298°K. The subscript aq appended to e$^-$ and H$^+$ indicates that protons and electrons are subatomic in size and strongly bound to water molecules. All species participating in solution reactions should, of course, be understood to be solvated. Many of the reactions listed are presumed to be elementary reactions, while others will presumably be shown to be complex processes sooner or later.

157

Entropy of activation values are interpreted in terms of the relative ordered-ness of the structures of the activated complex and the reactants. A positive entropy of activation would imply that the activated complex structure is less ordered than the structure of the separated reactants, whereas a negative entropy of activation (the usual case) would imply that the activated complex structure is more ordered than the structure of the separated reactants. If we compare the thermodynamic formulation of the ACT rate constant for bimolecular reactions with the line-of-centers rate constant formula, we notice that the factor $\exp(\Delta S^{\dagger}/R)$ plays a role similar to the steric factor p. A low value of p and a negative value of $\Delta S^{\dagger}$ therefore have essentially the same meaning.

Many applications of the thermodynamic formulation of ACT to the interpretation of solution reactions are given in the Supplementary References at the end of this chapter. An example of the computational procedure that is required is given in Exercise 6-13.

6-2 IONIC REACTION MECHANISMS AND THE PRIMARY SALT EFFECT

Before discussing elementary reactions involving ions, we need to review briefly the thermodynamic treatment of ionic equilibrium. Solution equilibria are governed by sets of equations of the form

$$\exp(-\Delta G^0/RT) = \prod a_i^{\nu_i} \equiv K_a = \prod ([A_i]\gamma_i)^{\nu_i} = \prod [A_i]^{\nu_i} \prod \gamma_i^{\nu_i} \equiv K_c K_\gamma$$

which relate standard free energy changes to *activity* products $\prod a_i^{\nu_i}$, denoted K_a, which are connected in turn to equilibrium constants in terms of concentrations by $K_a = K_c K_\gamma$, with the *activity coefficient* γ_i defined as the ratio $a_i/[A_i]$.

If K_γ were known, then there would be a simple relationship connecting ΔG^0 with K_c again, and we could return to the thermodynamics used heretofore. Unfortunately, activity coefficients for individual ions are only known for dilute solutions. For aqueous solutions at 25°C, there is an approximate theory (the Debye–Hückel theory) which predicts that γ_i depends on the ionic charge z_i and the ionic strength $I \equiv \frac{1}{2} \sum m_i z_i^2$, where $m_i \equiv n_i/1000$ g solvent, according to

$$\gamma_i = \exp(-1.17z_i^2 I^{\frac{1}{2}})$$

The summation appearing in the definition of the ionic strength I must be taken over all species in the solution, whether or not they are involved in the equilibrium being considered. If this theoretical relationship is extrapolated upwards from the concentration range where the assumptions made in its derivation are valid into the concentration range of most equilibrium and kinetics measurements, then we have a means of computing K_γ values. Extrapolation of equations is always risky, but scientists routinely do so either in order to make new discoveries about the limitations of the theories underlying the equation being extrapolated, or for lack of insight into the correct equation to use, or because the more correct equations are too complex mathematically or contain too many uncertain parameters. In the instance we deal with now, we have to combine the approximate γ_i equation with the activated complex theory, whose uncertainties were discussed in Chapter 4; nevertheless results in good agreement with experiment will be obtained.

We reconsider the postulated equilibrium between reactants A and B and activated complexes C

$$A + B \text{ "}\rightleftharpoons\text{" } C$$

for the thermodynamics appropriate for ionic solutions, and then postulate as in Section 6-1 that the rate of reaction is proportional to the concentration of activated complexes.

Since K_c is not a true constant, we consider the K_a expression

$$K_a \equiv \frac{a_c}{a_A a_B} = \exp(-\Delta G^{0\dagger}/RT)$$

as governing the postulated equilibrium, then relate activities to concentrations by $K_a = K_c K_\gamma$. A few algebraic maneuvers yields for the concentration of activated complexes

$$[C] = K_\gamma^{-1} \exp(-\Delta G^{0\dagger}/RT)[A][B]$$

and, recalling the procedures of ACT used in Section 6-1, for the rate of the elementary step $A + B \rightarrow$ products

$$R = \frac{kT}{h} K_\gamma^{-1} \exp(-\Delta G^{0\dagger}/RT)[A][B]$$

Finally, making use of the activity coefficient equation for aqueous ionic solutions to evaluate K_γ,

$$R = \frac{kT}{h} \exp(2.34 z_A z_B I^{\frac{1}{2}}) \exp(-G^{0\dagger}/RT)[A][B]$$

We have recovered the thermodynamic ACT formula, corrected for the effect of ionic strength by inclusion of the factor $\exp(2.34 z_A z_B I^{\frac{1}{2}})$.

EXERCISE 6-3 Write the complete derivation outlined in the foregoing paragraph.   You will need to make use of the fact that $z_C = z_A + z_B$.

There are several ways of comparing this result with experiments on reaction rates in ionic solutions, one of which is to draw semilogarithmic graphs of k/k_0 versus $I^{\frac{1}{2}}$, where k_0 is the rate constant extrapolated to zero ionic strength.   Our theory predicts the graphs shown in Figure 6-1.   If

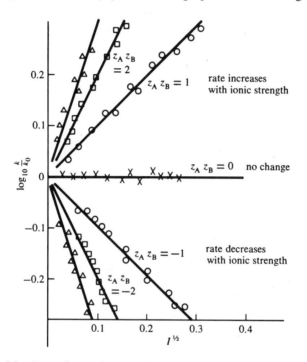

FIGURE 6-1   *Dependence of measured second order rate constants on ionic strength for various values of the ion charge product $z_A z_B$.   The data points represent a typical degree of agreement between theory and experiment for the multitude of reactions with which the primary salt effect has been studied. The rate constant at zero ionic strength k_0 is actually a limiting value for low ionic strength when $z_A z_B \neq 0$; since the theory only predicts the ratio k/k_0 , the actual value of k_0 is unimportant for testing the theory.*

Some examples of ionic reactions with which the primary salt effect has been studied are

$$S_2O_8{}^{2-} + 2I^- \rightarrow I_2 + 2SO_4{}^{2-} \quad (z_A z_B = 2)$$

$$O_2NNCOOC_2H_5{}^- + OH^- \rightarrow N_2O + CO_3{}^{2-} + C_2H_5OH \quad (z_A z_B = 1)$$

$$H_2O_2 + 2H^+ + 2Br^- \rightarrow 2H_2O + Br_2 \quad (z_A z_B = -1)$$

$$Co(NH_3)_5Br^{2+} + OH^- \rightarrow Co(NH_3)_5OH^{2+} + Br^- \quad (z_A z_B = -2)$$

either A or B, or both, are uncharged, the rate constant is independent of I; otherwise the logarithm of the rate constant either increases ($z_A z_B > 0$) or decreases ($z_A z_B < 0$) linearly with $I^{\frac{1}{2}}$ as shown. The data points indicate roughly the correlation found experimentally between measured conversion rates and $I^{\frac{1}{2}}$, with $z_A z_B$ providing an excellent clue for identifying the elementary step that is rate controlling.

The ionic strength influence on a reaction rate is called the *primary salt effect*. Note that it is a large effect: for $z_A z_B = 2$, $k = 1.6 \, k_0$ at $I^{\frac{1}{2}} = 0.1$, which would correspond to a 0.01-molar solution of NaCl. The salt effect could be regarded a form of catalysis, since a conversion rate is altered by addition of a substance that does not appear in the stoichiometric equation. However, it is probably better not to regard it as catalysis. The salt really affects the environment in which the reaction occurs by changing the strength of the interionic electrostatic forces rather than by changing the reaction mechanism.

6-3 ACID–BASE CATALYSIS

Solution reactions having mechanisms involving proton transfers are common. The mechanisms have been studied thoroughly with particular regard to the catalytic influence of acids and bases upon the reaction rates. It is customary in this field of study to use the Brönsted–Lowry definitions of acids and bases, according to which a substance that donates a proton, as $HA \rightarrow H^+ + A^-$, is acting as an acid, and a substance that accepts a proton, as $A^- + H^+ \rightarrow HA$, is acting as a base. It is also customary to take into account the *secondary salt effect*, which is the effect of ionic strength on acid and base concentrations owing to the interconnection between activity coefficients and ionic strength discussed in the preceding section.

We consider reactions that are first order in a reactant S, which is called the *substrate*. Substrate molecules may enter into elementary reactions with acid or base molecules in the catalyzed mechanism; the first order rate constant in $R \equiv k_{cat} v_S^{-1} [S]$ is then written

$$k_{cat} = k_0 + k_{H^+}[H^+] + k_{OH^-}[OH^-] + k_{HA}[HA] + k_{A^-}[A^-]$$

where k_0 now refers to the uncatalyzed rate, and k_{H^+}, k_{OH^-}, k_{HA}, and k_{A^-} are *catalytic constants*. If k_{HA} is important, the reaction is said to be subject to *general acid catalysis*; if k_{A^-} is important, to *general base catalysis*. If only k_{H^+} and/or k_{OH^-} are important, the reaction is said to be subject to *specific acid catalysis* and/or *specific base catalysis*.

This formulation of k_{cat} in terms of four separate catalysts should be regarded only as a formalism. In a real solution, these concentrations are coupled by equilibrium relationships, and the connection between their values and the catalytic mechanism may be very indirect.

To measure the catalytic constants for specific acid and/or base catalysis, R is measured in buffered solutions over as wide a pH range as possible, with care taken to maintain constant ionic strength. Since $[H^+]$ and $[OH^-]$ are related to one another at 298°K by the ion product constant $[H^+][OH^-] = 1.0 \times 10^{-14}$ mole² liter^{-2}, their ratio can be varied by a factor of 10^{12} without using acid or base concentrations greater than 0.1 molar. This large factor means that k_{H^+} and k_{OH^-} can be of quite different magnitudes for a given reaction and still both be measurable. Interpretation of the results in terms of mechanisms may be straightforward, but more often than not it is necessary to be careful in deciding upon elementary reactions that produce the observed catalysis. Examples are given in the exercises at the end of the chapter.

The general procedure for determining k_{HA} and k_{A^-} values is simple. One buffers the solution at pH values where neither $k_{H^+}[H^+]$ nor $k_{OH^-}[OH^-]$ overwhelms the contributions of $k_{HA}[HA]$ and/or $k_{A^-}[A^-]$ and maintains constant ionic strength by addition of a neutral salt. Then the conversion rate is measured at two different pH values as a function of $[HA]$, for example. At a given pH the ratio $[HA]/[A^-]$ is constant. The slopes of the two resulting k_{cat} versus $[HA]$ curves give two equations from which k_{HA} and k_{A^-} can be calculated.

An interesting relationship between k_{cat} for reactions subject to general acid or base catalysis and the dissociation constant of the catalyst acid $K_a \equiv [H^+][A^-]/[HA]$ was discovered by Brönsted. (Heretofore K_a has denoted an equilibrium constant in terms of *activities*; hereafter K_a denotes an acid dissociation constant.) For catalysis by monobasic acids, the relationship is

$$k_{cat} = \beta K_a{}^\alpha$$

If there are p separate groups on the catalyst acid that can yield protons, and q different atoms in the conjugate base that can accept protons, the relationships are

$$\frac{k_{cat}}{p} = \beta \left(\frac{q}{p} K_a \right)^\alpha$$

for general acid catalysis and

$$\frac{k_{cat}}{p} = \beta \left(\frac{p}{q} K_a \right)^\alpha$$

for general base catalysis. Such "Brönsted relationships" have been found to hold for many reactions. For α near zero, the solvent is the principal catalyst; for α near unity, H^+ is the principal catalyst; intermediate values correspond to general catalysis. An approximate theoretical basis for these relationships is presented in Section 6-4.

We return to the special subject of protonation and deprotonation reactions in Section 6-5.

6-4 LINEAR FREE ENERGY RELATIONSHIPS

The Brönsted relationships are special cases of *linear free energy* (LFE) *relationships*, which play a prominent role in the interpretation of elementary steps of organic reactions. Such relationships have been studied for reactions having ionic, radical, or molecular mechanisms, in short for all kinds of solution reactions. They are based on a judicious combination of thermodynamic and structural ideas, which we introduce here in terms of an atom transfer step.

A qualitative graph of potential energy versus distance for the transfer of atom from X to Y or to a molecule Y' similar to Y

$$XA + Y \rightarrow X + YA$$

$$XA + Y \rightarrow X + Y'A$$

is shown in Figure 6-2. The XA, YA, and Y'A curves are presumed to be almost *linear* near the transition state X-A-Y, and the difference between the potential energy curves for YA and Y'A in this region is presumed to be a *vertical* displacement. The variation in activation energy δE_A when the rates for various atom acceptors Y' are compared with the Y rate is then, by the geometry of the graphs, a constant fraction α of the variation in the energy change $\delta \Delta E^0$, giving $\delta E_A = \alpha \delta \Delta E^0$. If, for the reaction concerned, the ΔpV and ΔTS terms are the same for Y and Y' acceptors, then, from $G \equiv E + pV - TS$, it is also true that $\delta G^\dagger = \alpha \delta \Delta G^0$. The rate constant ratio, assuming that the thermodynamic formulation of ACT (Section 6-1) is applicable, is then

$$\frac{k'}{k} = \frac{\exp(-\Delta G^{\dagger'}/RT)}{\exp(-\Delta G^\dagger/RT)} = \frac{\exp(-\alpha \Delta G^{0'}/RT)}{\exp(-\alpha \Delta G^0/RT)} = \left(\frac{K'}{K}\right)^\alpha$$

where K and K' are equilibrium constants for the reactions $XA + Y \rightleftharpoons X + YA$ and $XA + Y' \rightleftharpoons X + Y'A$, respectively. This equation predicts that the rate constants for a series of related reactions (similar Y's) can be calculated from the differences in ΔG^0 values. In practice, a reference reaction is chosen, and the applicability of the LFE idea to this series of reactions is tested by plotting $\delta \log k$ versus $\delta \log K$.

EXERCISE 6-4 Write out in detail the derivation given in the preceding paragraph.

In view of the many assumptions made in deriving the LFE equation, one might expect that it would apply only to a small number of reactions. On the contrary: the correlations that are found prove to be among the most accurate and most general ones in all of chemical kinetics. There appears to be a natural law that the conflicting parameters that one would think should affect relative rate constants in addition to $\delta \Delta G^0$ always average out the same way. When exceptions to LFE relationships appear, there are usually plausible explanations to account for them.

The best known of all LFE relationships is the Hammett $\rho\sigma$ equation, which relates the effects of *meta-* or *para-*substituents on the rate constants, and

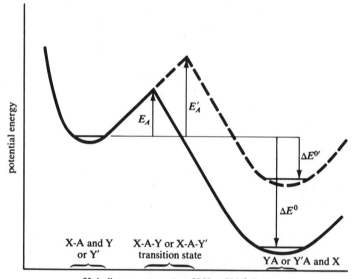

X-A distance at constant X-Y or X-Y' distance

FIGURE 6-2 *Graphs of energy versus reaction coordinate for two atom transfer steps* X-A + Y → X + YA *(solid line) and* X-A + Y' → X + Y'A *(dashed line). The dashed product curve is a vertical displacement of the solid product curve. Regions on the reaction coordinate identifiable with reactant states, transition states, and product states are indicated. The variation in activation energy between solid and dashed curves* $\delta E_A \equiv E_A' - E_A$ *is related to the variation in energy minima between solid and dashed curves* $\delta \Delta E^0 \equiv \Delta E^{0\prime} - \Delta E^0$ *by the equation* $\delta E = \alpha \delta \Delta E^0$*, where* α *is related to the slopes of the graphs (cf. Exercise 6-7). Remember that the term " energy " is used very loosely in graphs of this kind: thermodynamic, potential, and activation energies are indiscriminately intercompared. A more careful comparison is given in Figure 3-1.*

equilibrium constants, for reactions involving aromatic compounds. The reference reaction is the dissociation of benzoic acid in water at 298°K

$$C_6H_5COOH \overset{K_a^0}{\rightleftharpoons} C_6H_5COO^- + H^+$$

$K_a^0 = 10^{-4.20}$ moles/liter

A number σ characteristic of substituent X is found from its acid dissociation in water at 298°K

$$XC_6H_5COOH \overset{K_a^X}{\rightleftharpoons} XC_6H_5COO^- + H^+$$

The definition is $\sigma \equiv \log(K_a^X/K^0)$. The Hammett equations

$$\log(k_X/k_0) = \rho\sigma$$

$$\log(K^X/K^0) = \rho\sigma$$

then define rate or equilibrium constant ratios for reactions that the substituted compounds undergo. The constant ρ characterizes the reaction of interest, the constant σ the substituent. Tables of σ and ρ data can be found in advanced textbooks of organic chemistry.

We must abruptly discontinue this start of an excursion into physical organic chemistry, for it is a vast topic that prerequires considerable knowledge of descriptive organic chemistry for further exposition. As a student of chemical kinetics you need to know that there exist highly successful relationships, whose theoretical foundations are uncertain, correlating rate and equilibrium constants of groups of organic reactions having the same mechanisms.

An example of the use of the Hammett equation is given in Exercise 6-14.

6-5 DIFFUSION CONTROLLED ELEMENTARY STEPS OF SOLUTION REACTIONS

In Chapter 4 we considered the notion that an upper limit to the rates of bimolecular steps of gas reactions is set by the number of encounters per second between potentially reacting molecules. There is an analogous upper limit for bimolecular steps of solution reactions. The solution analog of the hard sphere encounter frequency of the kinetic theory of gases was derived in 1917 by M. V. Smoluchowski. He assumed that the solute molecules behave like spheres undergoing Brownian motion in a viscous fluid and showed that

$$^1R_E = 4\pi(D_A + D_B)(r_A + r_B)\,^1N_A\,^1N_B$$

^1R_E is the number of encounters per cm^3 per second, ^1N_A and ^1N_B are the concentrations of A and B in molecules per cm^3, r_A and r_B are the radii of the molecules, and D_A and D_B are the diffusion coefficients of A and B. The origin of this ^1R_E expression can be understood by considering the following simplified derivation.

The flux per second of A molecules through a unit area is proportional to the concentration gradient, with the constant of proportionality being the diffusion coefficient D_A:

$$\text{flux (per } cm^2 \text{ per sec)} = -D_A\frac{d\,^1N_A}{dx}$$

Consider the total flux inward through the surface of a sphere of radius r, centered at a B molecule, when the concentration gradient is radial. Since the surface area is $4\pi r^2$, the total flux into the sphere is

$$R_B \equiv \text{flux at distance } r \text{ from B} = 4\pi r^2 D_A\frac{d\,^1N_A}{dr}$$

Assume that when an A molecule diffuses as close as $(r_A + r_B)$ it reacts, so that ^1N_A at $r = (r_A + r_B)$ is 0. At large r, ^1N_A has the bulk concentration value. The flux to one B molecule is its rate of diffusion controlled reaction with A molecules. Integrating between the stated limits gives

$$R_B\int_{(r_A+r_B)}^{\infty}\frac{dr}{r^2} = 4\pi D_A\int_0^{\,^1N_A} d\,^1N_A$$

$$R_B = 4\pi(r_A + r_B)D_A\,^1N_A$$

The fact that the B molecule is not stationary, but undergoes Brownian motion as well, may be taken into account by replacing D_A with $(D_A + D_B)$. Multiplying R_B by ^1N_B gives the total diffusive encounter rate ^1R_E

$$^1R_E = 4\pi(r_A + r_B)(D_A + D_B)\,^1N_A\,^1N_B$$

We can convert this immediately to a second order rate constant in our usual units of liters mole^{-1} sec^{-1}

$$k_{diff} = \frac{4\pi \cdot 6 \cdot 10^{23}(D_A + D_B)(r_A + r_B)}{1000}$$

This equation gives a false impression of the dependence of k_{diff} on molecular parameters, for D_A and D_B are inversely proportional to r_A and r_B. For the same solution model used by Smoluchowski, the Stokes–Einstein equation $D = kT/6\pi r\eta(T)$ holds, and the k_{diff} equation can be recast

$$k_{diff} = \frac{2RT}{3000\eta(T)}\frac{(r_A + r_B)^2}{r_A r_B}$$

where $\eta(T)$ is the viscosity of the fluid in poise. For the case in which $r_A = r_B$, the simple equation

$$k_{diff} = \frac{8RT}{3000\eta(T)}$$

results.

The temperature dependence of $\eta(T)$ can be written in Arrhenius form. The resulting activation energies for viscosity are found to be about $\frac{1}{3}$ of the latent heats of vaporization of the solvents concerned. For water the temperature dependence of k_{diff} in the above equation is found from experimental viscosity data to correspond to an activation energy of $298 < T < 308°K$. For organic solvents E_A is typically 2–3 kcal.

EXERCISE 6-5 Verify the above E_A result from the viscosity data for water: $\eta(298°) = 0.8937$ centipoise, $\eta(308°) = 0.7225$ centipoise. Incidentally, notice that whereas the rates of solution reactions typically double for a 10° temperature increase, the resistance to flow only decreases by about 20% for the 10° temperature increase here.

We conclude that diffusion controlled reactions are characterized by low activation energies. This can be interpreted as arising from the fact that the potential energy barriers that reactant molecules must overcome in diffusing through solvent molecules are small compared with the potential energy barriers usually encountered on reaction coordinates. The simple final equation for k_{diff} is a very useful means of estimating the upper limit for solution reaction rates. It gives a numerical value for k_{diff} in water at

298°K of 7.4×10^9 liters mole^{-1} sec^{-1}, which indeed proves to be of the magnitude of the fastest rate constants for reactions not involving ions.

Two factors complicate the situation when ions are involved. First, if both reactants are ions, then the electrostatic forces between them must be taken into account. Then the k_{diff} equations become

$$k_{diff} = \frac{4\pi \cdot 6 \cdot 10^{23}(D_A + D_B)(r_A + r_B)}{1000} \frac{\delta}{\exp(\delta) - 1}$$

or

$$k_{diff} = \frac{8RT}{3000\eta(T)} \frac{\delta}{\exp(\delta) - 1}$$

where

$$\delta \equiv z_A z_B e^2 / D(r_A + r_B)kT$$

where z_A and z_B are the ionic charges, e is the value of the electronic charge, 1.6×10^{-19} coulombs, D is the dielectric constant of the solvent, and k is, as usual, the Boltzmann constant. For an elementary step in water between a univalent cation and a univalent anion, the correction factor to k_{diff} may range from 1 to 10, depending on the value of $(r_A + r_B)$. For ion–ion reactions, diffusion controlled rates will, therefore, have rate constants in the range 10^{10}–10^{11} liters mole^{-1} sec^{-1}. A second complication is that ions can interact strongly with the solvent cages surrounding them, which is incompatible with the assumptions made in deriving the formulas for k_{diff}. This effect, which may also be important for reactions of neutral molecules, requires rather tenuous and approximate theories for its interpretation.

The maximum rate of a unimolecular dissociation in solution is determined by the rate at which the products can diffuse away from the solvent cage in which they are formed. This rate can be found by the same procedure developed above if boundary conditions (limits on the integrals) appropriate to a diffusion controlled dissociation are used (cf. Exercise 6-19). If the products are uncharged and do not interact with one another, the resulting equations are

$$k_{diff}^1 = \frac{3(D_A + D_B)}{(r_A + r_B)^2}$$

and for $r_A = r_B = r_{AB}/2$

$$k_{diff}^1 = \frac{2kT}{\pi\eta(T)r_{AB}}$$

If the products are ions that interact by Coulomb forces, these equations become

$$k_{diff}^1 = \frac{3(D_A + D_B)}{(r_A + r_B)^2} \left[\frac{\delta}{1 - \exp(-\delta)}\right]$$

and

$$k_{diff}^1 = \frac{2kT}{\pi\eta(T)r_{AB}} \left[\frac{\delta}{1 - \exp(-\delta)}\right]$$

Reactions of H^+ and OH^- ions, called *protonation and deprotonation reactions*, respectively, prove to be diffusion controlled, with k values in the range 10^{10}–10^{11} liters mole^{-1} sec^{-1}, unless unusual charge or steric effects are important. Diffusion controlled rates also appear in proton transfer reactions. If DH^+ is a proton donor and A is a proton acceptor, the simplest proton transfer mechanism is

$$DH^+ + A \underset{k_{-1}}{\overset{k_1}{\rightleftharpoons}} D—H^+—A \underset{k_{-2}}{\overset{k_2}{\rightleftharpoons}} D + HA^+$$

Formation of the intermediate $D—H^+—A$ is assumed to be diffusion controlled from either direction; that is, $k_1 \cong k_{-2} \cong 10^{10}$–$10^{11}$. Applying the steady state approximation to $D—H^+—A$ yields

$$k_f = \frac{k_1}{1 + k_{-1}/k_2} \qquad k_r = \frac{k_{-2}}{1 + k_2/k_{-1}}$$

The ratio k_{-1}/k_2 is determined by the relative stabilities of DH^+ and HA^+; if DH^+ is more stable than HA^+, then $D—H^+—A$ will decompose most frequently to give $DH^+ + A$, so that $k_{-1}/k_2 \gg 1$ and $k_f = k_1 k_2/k_{-1}$; then $k_r = k_{-2}$ is diffusion controlled. For the converse case, in which HA^+ is more stable than DH^+, $k_{-1}/k_2 \ll 1$, thus giving $k_r = k_{-2}k_{-1}/k_2$, and $k_f = k_1$ is diffusion controlled. In short, proton transfer reactions are diffusion controlled in the thermodynamically favored direction and controlled in the unfavored direction by the (far slower) rate of bond rupture.

EXERCISE 6-6 Write out in detail the argument of the preceding paragraph.

As a concluding remark on diffusion controlled reactions, let us observe that if the diffusion coefficients D_A and D_B are known, or estimated, then Smoluchowski's original equation for k_{diff} can be used to calculate a characteristic distance $r_{AB} \equiv (r_A + r_B)$ from any experimental k_{diff} value. This is the solution reaction counterpart to the reactive collision diameter σ_R introduced in Chapter 4 for bimolecular gas reactions. The value of r_{AB} can have similar use in interpreting the transition state geometry.

EXERCISES

6-7   Write the equations for the straight lines of the graphs in Figure 6-2, and rearrange them to find the relationship between α and the slopes of the lines. What limiting values for α are imposed by geometry?

6-8   The reaction 2 cyclopentadiene $\rightarrow$ endodicyclopentadiene has a mechanism identical to its stoichiometric equation in the gas phase and in a variety of solvents. The Arrhenius parameters in the gas phase and in CS_2 solution are $E_A = 16.8$ kcal and $\log_{10} A = 6.2$.   (a) Calculate $t_{\frac{1}{2}}$ for $[\text{cyclopentadiene}]_0 = 2.82$ moles/liter, $T = 273°K$.   (b) Calculate σ_R (273) and p in the line-of-centers formula.   (c) Calculate the number of collisions that an "average" cyclopentadiene molecule in a CS_2 solution of $[\text{cyclopentadiene}]_0 = 2.82$ moles/liter will have with (1) other cyclopentadiene molecules and (2) CS_2 molecules before finally reacting.   Assume both σ_{AB} values to be 10 Å^2. Density of $CS_2 = 1.3$ g/cc.

6-9   In concentrated $HClO_4$ solution the rate law for the reaction

$$2Co^{3+} + H_2O_2 \rightarrow 2Co^{2+} + 2H^+ + O_2$$

was found to be $R = k[Co^{+3}]^{1.0}[H_2O_2]^{1.0}[H^+]^{-1.0}$.   (a) Propose a mechanism for this reaction with HO_2^- and HO_2 as intermediates.   The acid dissociation constant K_a for H_2O_2 is 2×10^{-12} moles/liter.   (b) Draw a graph of k/k_0 versus $I^{\frac{1}{2}}$ consistent with your mechanism.

6-10   Consider an acid catalyzed reaction having the mechanism

$$HA \rightleftharpoons H^+ + A^-$$

$$HA + S \rightleftharpoons SH^+ + A^-$$

$$SH^+ + H_2O \rightarrow \text{products}$$

(a) Derive the rate law for this mechanism by applying the steady state assumption to the intermediate SH^+.   (b) Show that this rate law has two limiting forms, corresponding to general and specific acid catalysis, respectively.

6-11   The acid catalysis of the dehydration reaction of acetaldehyde hydrate (cf. Section 2-4) has been studied for many catalysts.   Calculate the value of α in the Brönsted relationship for this reaction from the data: k_{cat} for acetic acid $= 0.32$ liters mole^{-1} sec^{-1}; K_a for acetic acid $= 1.8 \times 10^{-5}$ moles/liter; k_{cat} for chloroacetic acid $= 2.4$ liters mole^{-1} sec^{-1}; K_a for chloroacetic acid $= 1.5 \times 10^{-3}$ moles/liter.   Does this result indicate H^+ catalysis, or rather HA catalysis?

6-12   Anthracene dissolved in benzene fluoresces when irradiated with uv light (cf. Section 10-2).   The fluorescence is "quenched" (i.e., suppressed) by dissolved oxygen because of the elementary step

$$A^T + O_2 \rightarrow A + O_2 + \text{kinetic energy}$$

which converts the potentially fluorescing A^T molecules into ordinary anthracene molecules A. The experimental value of the second order rate constant in $R \equiv k[A^T][O_2]$ is 3.0×10^{11} liters mole^{-1} sec^{-1} at 298°K, and the experimental diffusion coefficients are D_A (assumed $= D_A{}^T) = 2.2 \times 10^{-5}$ and $D_{O_2} = 5.7 \times 10^{-5}$ cm^2 sec^{-1}. (a) Calculate the reaction distance $(r_A + r_B)$, assuming the quenching to be diffusion controlled. (b) Compare the experimental k with $8RT/3000\eta(T)$, and account for the discrepancy. $\eta(298)$ for benzene $= 0.0060$ poise (cf. Section 10-2e).

6-13 Determination and interpretation of $\Delta S^\dagger$ values is a major activity of physical organic chemists. As a sample of this, consider the basic hydrolyses (saponifications) of aliphatic esters. (a) Calculate $\Delta S^\dagger$ values for the saponifications of ethyl acetate, ethyl propionate, and ethyl butyrate from the following data. $k_{Ac}(293) = 4.5 \times 10^{-3}$ liters mole^{-1} sec^{-1}, $E_A = 14.7$ kcal; $k_{Prop}(293) = 2.3 \times 10^{-3}$ liters mole^{-1} sec^{-1}, $E_A = 14.7$ kcal; $k_{But}(293) = 1.2 \times 10^{-3}$ liters mole^{-1} sec^{-1}, $E_A = 15.1$ kcal. (b) On the basis of the (simplified) arguments in Section 6-3, would this set be expected to obey a linear free energy relationship?

6-14 The acid dissociation constants of p-NO$_2$ and m-NO$_2$ benzoic acids are $10^{-3.42}$ and $10^{-3.49}$ moles/liter, respectively. The saponification of p-NO$_2$ ethyl benzoate in 85% ethyl alcohol solvent at 298°K was found to be faster than the saponification of ethyl benzoate itself under the same conditions by a factor of 116. Calculate the second order rate constant for m-NO$_2$ ethyl benzoate saponification from the p-NO$_2$ ethyl benzoate rate constant, $k(298) = 7.2 \times 10^{-4}$ liters mole^{-1} sec^{-1}, and the Hammett $\rho\sigma$ equation.

6-15 The thermodynamic formulation of ACT can be used to interpret the effect of pressure on solution reaction rates. (a) Find $(\partial \bar{G}^0/\partial p)_T$, $(\partial \Delta G^0/\partial p)_T$, $(\partial \Delta G^{0\dagger}/\partial p)_T$, and $(\partial \ln K^\dagger/\partial p)_T$ in terms of volume changes and temperature. (b) Sketch a semilogarithmic graph of k versus p. (c) Suggest a possible interpretation of the "volume of activation" $\Delta \bar{V}^\dagger$ in your expression for $(\partial \ln K^\dagger/\partial p)_T$.

6-16 Why can one not obtain a valid general expression for a bimolecular reaction rate in solution by multiplying the hard sphere encounter frequency ^1R_E by the fraction of encounters involving an energy greater than the activation energy, $\exp(-E_A/RT)$, in the manner employed in the collision theory of bimolecular reaction rates? (Compare with remarks on page 151. Is an encounter the same as a collision? Could there be more than one collision between reactants in a single encounter?)

6-17 By an extended treatment of the boundary conditions for diffusion controlled reaction rates it can be shown that the result obtained in Section 6-4 for k_{diff} of uncharged reactants should be divided by $1 + [4\pi(D_A + D_B)(r_A + r_B)(1 - p/2)] \times Z_{AB}^{-1} p^{-1}$, where Z_{AB} is the encounter frequency for A and B and p is the probability of A and B reacting at an encounter. Find the $p \ll 1$ and $p = 1$ limiting cases for the corrected k_{diff} formula.

6-18 Diffusion controlled reactions have rate constants near 10^{10} liters mole^{-1} sec^{-1}. How would you reconcile this fact with the A factors of Table 6-1 that greatly exceed this value?

6-19 The limits for the integrals that must be evaluated in calculating the rate of diffusion controlled unimolecular dissociation are $^1N_A = 0$ for $r = \infty$ and $^1N_A = [4(r_A + r_B)^3/3]^{-1}$ for $r = r_A + r_B$. Using the procedure given in Section 6-5 for bimolecular reactions, derive the expressions for k_{diff} for unimolecular decomposition giving uncharged, noninteracting products.

6-20   Find an expression for the entropy change in a dissociation–association reaction $A + B \rightleftharpoons AB$ by forming the ratio of the diffusion controlled forward and reverse rate constants, and relating this equilibrium constant to a standard free energy change.

6-21   Propose mechanisms for the four chemical reactions given in the caption to Figure 6-1 that are consistent with the primary salt effects that are observed in each case.

6-22   The catalytic rate constant for oxygen isotopic exchange between acetone and water in acidic solution at 298°K can be written in the form $k_{cat} = k_H[H^+] + k_{HA}[HA] + k_A[A^-]$. Determine k_H, k_{HA}, and k_A from the following data:

[HA]/[A⁻]	0.2	0.2	1.0	1.0
$[HA]/[A^-]$	0.2	0.2	1.0	1.0
k_{cat}, sec⁻¹	1.5×10^4	2.7×10^4	4.3×10^4	4.8×10^4
[HA], moles/liter	0.0225	0.100	0.0135	0.050

SUPPLEMENTARY REFERENCES

General books on reactions in solution are

A. A. Frost and R. G. Pearson, *Kinetics and Mechanism* (J. Wiley & Sons, Inc., New York, 1961).

E. A. Moelwyn-Hughes, *The Kinetics of Reactions in Solution* (Oxford University Press, Oxford, 1947).

Acid–base catalysis is discussed by

R. P. Bell, *The Proton in Chemistry* (Cornell University Press, Ithaca, N.Y., 1959).

Linear free energy relationships are treated by

J. E. Leffler and E. Grunwald, *Rates and Equilibria of Organic Reactions* (J. Wiley & Sons, Inc., New York, 1963).

L. P. Hammett, *Physical Organic Chemistry* (McGraw-Hill, New York, 1940).

P. R. Wells, *Linear Free Energy Relationships* (Academic Press, New York, 1968).

The literature of mechanism studies on solution reactions is large. Some books primarily on organic reactions are

R. Breslow, *Organic Reaction Mechanisms* (W. A. Benjamin, Inc., New York, 1965).

E. S. Gould, *Mechanism and Structure in Organic Chemistry* (Holt, Rinehart and Winston, New York, 1963).

J. Hine, *Physical Organic Chemistry* (McGraw-Hill, New York, 1962).

C. K. Ingold, *Structure and Mechanism in Organic Chemistry* (Cornell University Press, Ithaca, N.Y., 1953).

P. Sykes, *A Guidebook to Mechanism in Organic Chemistry* (Longmans, London, 1961).

K. B. Wiberg, *Physical Organic Chemistry* (J. Wiley & Sons, Inc., New York, 1964), Part 3.

Some books primarily on inorganic reactions are

F. Basolo and R. G. Pearson, *Mechanisms of Inorganic Reactions* (J. Wiley & Sons, Inc., New York, 1967).

J. O. Edwards, *Inorganic Reaction Mechanisms* (W. A. Benjamin, Inc., New York, 1964).

J. Kleinberg, Ed., *Mechanisms of Inorganic Reactions; Advances in Chemistry Series 49* (American Chemical Society, Washington, 1965).

C. H. Langford and H. B. Gray, *Ligand Substitution Processes* (W. A. Benjamin, Inc., New York, 1965).

A compilation of solution reaction rate constants is published by the U.S. National Bureau of Standards:

Tables of Chemical Kinetics, Homogeneous Reactions, National Bureau of Standards Circular 510, 1951; Supplements 1 (1956) and 2 (1960); National Bureau of Standards Monograph 34 (1961).

Chapter 7

REACTIONS AT INTERFACES

WHEN A chemical reaction involves two different states of matter, as $Cu(s) + \frac{1}{2}O_2(g) \rightarrow CuO(s)$, then the reaction mechanism obviously includes some elementary reactions that take place at the boundary between them, which we call an *interface*. There are many additional reactions catalyzed through elementary reactions that occur at interfaces. Both cases are called *heterogeneous* reactions. In either case, we have to consider the role of the interface in the mechanism in a slightly different way than we have considered the reactant molecules of single-phase ("*homogeneous*") mechanisms. This is because the microscopic properties of the interfaces at which heterogeneous reactions occur are quite different from the microscopic properties of the same substances away from interfaces. The properties of interfaces have proved difficult to study.

The particular topics of catalysis of gas reactions at solid catalyst surfaces, electrode reactions, and enzyme reactions are of such great importance in chemistry that we should not omit discussing them. The presentations in this chapter, however, are but cursory introductions to these subjects. An entire book would be required to give an adequate introduction to the complex chemical kinetics of any of the three.

In practice it may be difficult to sort out elementary reactions that occur at an interface from the rest of a heterogeneous reaction mechanism. Frequently some of the elementary reactions are diffusion controlled either to, from, or at the interface, and can be identified by considerations similar to

those introduced in Section 6-5.   If a slow rate controlling step at the interface is involved, many experimental methods exist for obtaining information
about it.

7-1 HETEROGENEOUS CATALYSIS AT GAS–SOLID INTERFACES

Catalysis of gas reactions by solid surfaces plays a prominent part in industrial chemical processes.   Familiar examples are the Haber process for
nitrogen fixation by iron–nickel catalysts

$$N_2 + 3H_2 \underset{\longleftarrow}{\overset{metal}{\longrightarrow}} 2NH_3$$

cracking of hydrocarbons over silica–alumina catalysts

$$C_xH_y \xrightarrow[Al_2O_3]{SiO_2} \text{high octane gasoline}$$

and dehydrogenation of saturated hydrocarbons over metal oxide catalysts,
exemplified in a practical way by

$$\text{various hydrocarbons} \underset{oxide}{\overset{metal}{\rightleftarrows}} \text{butadiene} + \text{various other hydrocarbons}$$

Since ammonia becomes fertilizer, gasoline provides power, and butadiene
is a basic ingredient for the manufacture of tires, and since a host of other
valuable products can be made from inexpensive gases by catalytic methods,
the chemical industry pursues research on heterogeneous catalysis with great
vigor, protecting successful catalysts with blankets of patents and cloaks
of secrecy.

It is characteristic of heterogeneous catalysis experiments that the gap
between fundamental research on reaction mechanisms and practical research
on industrial catalysis is difficult to overcome.   In the fundamental research,
the simplest of reactions are studied using pure reagents and catalysts whose
purity is high and whose surfaces can be carefully characterized.   The reactions are carried out under carefully controlled conditions.   In practical
applications, however, none of these things can be realized: it is necessary to
deal with catalysts of commercial purity, with complex reactions, and with
commercial grade chemicals.   Furthermore, industrial processes require reactors in which the flow of gas over the catalyst is uneven and in which large
gradients of pressure and temperature exist.   The chemical kinetics of
heterogeneous reactions as introduced here, therefore, has only indirect
bearing on practical problems of heterogeneous catalysis.

The mechanism of a gas reaction at a solid surface can be subdivided into five parts:

(1) Movement of reactant molecules to surface (convection or diffusion)
(2) Adsorption of reactant molecules onto surface
(3) Elementary steps involving adsorbed molecules, either entirely on the surface or between adsorbed and gas phase molecules
(4) Desorption of product molecules from surface
(5) Removal of product molecules away from surface (convection or diffusion).

Any one of these can be rate limiting. Diffusion rates can sometimes be measured independently of reaction rates, allowing parts (1) and (5) of the mechanism to be clarified directly. Adsorption and desorption rates can frequently be studied independently of conversion rates, allowing parts (2) and (4) to be clarified.   Part (3), the heart of heterogeneous catalytic mechanisms, usually contains the rate controlling step.   The natures of the adsorbed species that react, the surface on which they are adsorbed, and the interactions of the adsorbed species with the surface and with each other are the principal objects of research on heterogeneous catalytic mechanisms.   Many kinds of indirect experiments and some direct ones are done to investigate them.

7-1a   Adsorption Isotherms

Insight into the interplay of adsorption and heterogeneous reaction was provided in 1916 by I. Langmuir, who derived several relationships between the amount of adsorption and the partial pressure of the adsorbed gas.   One of the simple cases he considered used the following model.   The molecules are assumed to be independently adsorbed on adsorption *sites*; at equilibrium at a given pressure, a fraction θ of the total number of sites has adsorbed molecules, while $(1 - \theta)$ of the sites are bare.   The rates of adsorption and desorption are equal at equilibrium.   The former is proportional to θ, the latter to $(1 - \theta)$ and to the pressure p.   Solving for θ as a function of p yields the Langmuir *adsorption isotherm*

$$\theta = \frac{bp}{1 + bp}$$

EXERCISE 7-1   (a) Write out a complete derivation of the Langmuir adsorption isotherm.   Find the low pressure and high pressure limiting forms, and draw a graph of θ versus p.   What kind of graph is predicted if θ^{-1} is plotted versus p^{-1}?   (b) If adsorption requires no activation energy, but desorption has activation energy $E_A = 6.5$ kcal, what will θ^{-1} versus p^{-1} graphs look like at $T = 300$ and $T = 310°K$? (c) What isotherm is predicted if the adsorption is very weak, that is, if $\theta \ll 1$?

EXERCISE 7-2 Suppose that the surface contains sites that can be occupied either by A molecules or by B molecules. Show that the isotherm giving θ_A as a function of p_A and p_B is $\theta_A = b_A\, p_A(1 + b_A\, p_A + b_B\, p_B)^{-1}$.

The Langmuir adsorption isotherm can also be derived by considering the adsorption process as an equilibrium between gas phase molecules A, adsorption sites S, and adsorbed molecules AS

$$A + S \rightleftarrows AS$$

Let ^1N_A be the number of A molecules per cubic centimeter, and let $^1N_{AS}$ and ^1N_S be the number of occupied and unoccupied sites, respectively, per square centimeter of surface. We then define an equilibrium constant for adsorption as

$$K_{ads} \equiv \frac{^1N_{AS}}{^1N_A\, ^1N_S}$$

The concentration of A molecules can be related to the pressure by the ideal gas law, giving

$$^1N_A = 6.02 \times 10^{23} p/RT = 7.34 \times 10^{21} p/T$$

if p is in atmospheres. The surface concentrations ^1N_S and $^1N_{AS}$ are related to the total number of occupied and unoccupied sites $^1N_{surface}$ by

$$^1N_{AS} = \theta\, ^1N_{surface}$$

$$^1N_S = (1 - \theta)\, ^1N_{surface}$$

Substituting these equations for the concentrations into the equilibrium constant expression gives

$$K_{ads} = \frac{\theta}{(1 - \theta)p} \cdot \frac{T}{7.34 \times 10^{21}}$$

This equation can be rearranged to give the Langmuir adsorption isotherm

$$\theta = \frac{bp}{1 + bp}$$

where b is now equal to $7.34 \times 10^{21} K_{ads}/T$ atm^{-1}.

This derivation has two advantages over the kinetic derivation given by Langmuir. In the first place it is independent of the actual mechanism of the adsorption process since it assumes only that adsorption can properly be considered as an equilibrium of the form $A + S \rightleftharpoons AS$. Second, we can use a statistical mechanical formula to evaluate K_{ads}. In Section 4-7 we used the statistical mechanical formula

$$K^\dagger = \frac{{}^1N_C}{{}^1N_A \, {}^1N_B} = \frac{q_C^*}{q_A^* q_B^*} \exp(-E_b/kT)$$

where the q^*'s are partition functions per unit volume, to describe the equilibrium $A + B \rightleftharpoons C$ between reactants and activated complexes. For the adsorption equilibrium the corresponding expression is

$$K_{ads} = \frac{{}^1N_{AS}}{{}^1N_A \, {}^1N_S} = \frac{q_{AS}^*}{q_A^* q_S^*} \exp(-E_{ads}/RT)$$

where q_A^* is the partition function per unit volume for the A molecules, and q_{AS}^* and q_S^* are partition functions per unit area for the occupied and unoccupied sites, respectively. The Boltzmann factor contains the heat of adsorption E_{ads}, which is a negative quantity, in contrast to the positive barrier height E_b appearing in the Boltzmann factor of $K^\dagger$. This reflects the fact that adsorption is exothermic, whereas formation of activated complexes is endothermic. The parameter b of the Langmuir adsorption isotherm is given by the statistical mechanical formula

$$b = \frac{7.34 \times 10^{21} q_{AS}}{T q_A \, q_S} \exp(-E_{ads}/RT)$$

The partition functions may be evaluated easily for the Langmuir model of the adsorption process. The surface is assumed to be rigid and has no internal quantum states; its partition function is therefore unity, or $q_S = 1$. The adsorbed molecules are assumed to be bound to individual sites and undergo vibrational motions there. It is plausible to expect that the fundamental frequency of vibration perpendicular to the surface v_z will be different from the frequency of the two vibrations parallel to the surface v_{xy}. Therefore the partition function for the adsorbed molecules is taken to be the product of harmonic oscillator partition functions

$$q_{AS} = [1 - \exp(-hv_z/kT)]^{-1}[1 - \exp(-hv_{xy}/kT)]^{-2}$$

The partition function per unit volume for the gas phase molecules A can be obtained (just as in ACT calculations, Section 4-7) either from tables of the free energy function or from rigid rotor, harmonic oscillator formulas. Let us here choose the latter course and factor the partition function per unit volume for A into an internal part $q_{A, int}$ and a translational part $q_{A, t} = (2\pi m_A kT/h^2)^{\frac{3}{2}}$. This gives

$$q_A = (2\pi m_A kT/h^2)^{\frac{3}{2}} q_{A, int}$$

(For adsorption of monatomic gases, $q_{A, int} = 1$.) Combining our statistical mechanical formulas gives for the parameter b (in atm^{-1}) the expression

$$b = \frac{7.34 \times 10^{21} \exp(-E_{ads}/RT)}{T[1 - \exp(-hv_z/kT)][1 - \exp(-hv_{xy}/kT)]^2 (2\pi m_A kT/h^2)^{\frac{3}{2}} q_{A, int}}$$

This equation shows that b, the inverse of the pressure at which half of the sites are occupied, is determined on the molecular scale by the heat of adsorption, the mass of A, the vibration frequencies at the adsorption site, and the internal partition function of A.

A flaw of the above derivation is that the heat of adsorption E_{ads} is considered to be independent of temperature and the same for all sites. It is found in calorimetric experiments, however, that heats of adsorption are temperature dependent and decrease as the degree of surface coverage increases. This can be attributed to the sites having different adsorption characteristics, that is, different values of b. If we divide the adsorption sites into groups according to their characteristic b_i values, where the subscript i indexes the groups, then the fraction of surface covered is given by the sum

$$\theta = \sum f_i \theta_i = \sum \frac{f_i b_i p}{1 + b_i p}$$

where f_i is the fraction of all sites belonging to group i. An important special case of variable heat of adsorption occurs if f_i is replaced by the distribution function

$$f(\varepsilon) = \alpha \exp(\alpha\varepsilon)$$

where α is the fraction of sites with energy of adsorption ε (still a negative quantity) between 0 and $d\varepsilon$. In this case the summation over groups of sites becomes the integral

$$\theta = \int_0^{-\infty} \frac{f(\varepsilon)b(\varepsilon)p}{1 + f(\varepsilon)b(\varepsilon)p} d\varepsilon$$

where $f(\varepsilon) = \alpha \exp(\alpha\varepsilon)$ and $b(\varepsilon)$ is given by the statistical mechanical formula with $E_{ads} = \varepsilon$. Evaluation of the integral (by sophisticated methods) gives

$$\theta = \alpha k T (b'p)^{\alpha k T}$$

which is of the same form as an empirical equation

$$\theta = ap^m$$

where a and m are experimental constants, called the *Freundlich isotherm*. It is found that adsorption data are usually in somewhat better agreement with the Freundlich isotherm than with the Langmuir isotherm, which can be interpreted as meaning that most surfaces have a distribution of adsorption sites.

7-1b   Rate Laws for Heterogeneous Catalysis

Several limiting cases of heterogeneous reaction kinetics can be interpreted in an enlightening way using the Langmuir adsorption isotherm.   Our fundamental assumption is that only adsorbed molecules can react.
   Case I: Suppose $R = k[A_{ads}]$ and A is weakly adsorbed.   For weak adsorption, $\theta = bp$.   (Exercise 7-1c.)   Therefore, $R = k'p_A$.
   Case II: Suppose $R = k[A_{ads}]$ and A is strongly adsorbed.   For strong adsorption, $\theta = 1$ (Exercise 7-1a.)   Therefore, $R = k''$,
   Case III: Suppose $R = k[A_{ads}]$, A is weakly adsorbed, and product B is strongly adsorbed.   For this case the isotherm gives $\theta_A = b_A\, p_A/(1 + b_B\, p_B)$, or, for strong adsorption of B, $b_A\, p_A/b_B\, p_B$.   (Exercise 7-2.)   Therefore, $R = k'p_A\, p_B^{-1}$.
   Numerous reactions have been found to follow these limiting cases, as well as the many easily envisaged intermediate ones.   As examples, the decomposition of NH_3 on platinum follows the rate law

$$R = k[NH_3]^{1.0}[H_2]^{-1.0}$$

and the rate of hydrogenation of ethylene on nickel was found to be given by

$$R = \frac{k[H_2]^{1.0}[C_2H_4]^{1.0}}{1 + b[C_2H_4]^{1.0}}$$

EXERCISE 7-3 Invent mechanisms for ammonia decomposition on platinum and hydrogenation of ethylene on nickel that are consistent with the rate laws given.

These examples would suggest, incorrectly, that heterogeneous catalysis is a subarea of chemical kinetics that is distinguished from homogeneous kinetics merely by the fact that mass transport and elementary reactions at an interface have to be included in a complete description of a heterogeneous reaction. To dispel this suggestion, we close this section with some cautionary remarks about surface phenomena. First, kinetics experiments on adsorption and desorption in the absence of reaction disclose that these processes themselves are actually complex. Writing an elementary reaction such as $A \rightarrow A_{ads}$ is a gross oversimplification of the mechanisms by which the great majority of gases interact with solid surfaces. In the presence of chemical reaction, the adsorption and desorption kinetics is likely to become far more complex still. Second, experiments on the nature of adsorbed species disclose that the designation "A_{ads}" is also a gross oversimplification of the diverse character of the physical and chemical bonding involved in adsorption. This diversity is characteristic not only of adsorption on chemically active catalysts, but also of adsorption on simple and carefully characterized surfaces such as single crystal faces or evaporated metal films. Finally, most heterogeneous gas–solid reactions yield many different products. The complex product distributions usually preclude straightforward interpretations of observed rate laws, and they may even preclude obtaining a satisfactory rate law at all. It is frequently the case in heterogeneous catalysis studies that one can do no more than suggest a plausible mechanism based on the product distribution and offer this mechanism, together with the empirical rate law for disappearance of reactant, as the entire scientific output of the research.

7-1c Maximum Rate of Adsorption

In Chapters 4 and 6 we noted that rate constants for homogeneous reactions have upper limits. In Section 4-2 we found that the maximum rate of a bimolecular gas reaction is limited by the encounter frequency, and in Section 6-4 we found that a bimolecular solution reaction could not have a rate greater than the rate at which reactant molecules diffuse to one another. Rate constants for heterogeneous reactions also have upper limits, which are set by the rate at which reactant molecules can reach the interface at which reaction takes place. We can derive an expression for the upper limit rate constant of a gas reaction catalyzed by a solid surface using the following simple model. Assume that the solid catalyst is in the form of small spheres of radius r_S dispersed uniformly in the reacting system at a concentration of ^1N_S spheres per cubic centimeter. The hard sphere encounter rate derived in Section 4-2 for A–BC collisions becomes for collisions of the molecules A with catalyst spheres S

$$^1R_E = (8\pi kT/m_A)^{\frac{1}{2}} r_S^2 \, ^1N_A \, ^1N_S$$

In this encounter rate expression the reduced mass μ has become m_A (since $m_S \gg m_A$, $\mu \equiv m_A m_S/(m_A + m_S) \cong m_A$) and σ^2 has become r_S^2 (since $r_S \gg r_A$). The total surface area $\mathscr{A}$ of the spheres in one cubic centimeter is $\mathscr{A} = \, ^1N_S \, 4\pi r_S^2$. If this area is factored from the encounter rate expression we obtain for the maximum rate at which A molecules can be absorbed on the surface

$$^1R_E = (2kT/\pi m_A)^{\frac{1}{2}} \mathscr{A} \, ^1N_A$$

The rate law for a gas reaction catalyzed at a solid surface of area $\mathscr{A}$ may be conveniently expressed (in molecules of A reacting per second) as

$$R = k_S \mathscr{A} \, ^1N_A$$

Comparing this rate law with the preceding relation for the rate at which A atoms arrive at the surface shows that the maximum value for the catalytic rate constant k_S is

$$\boxed{k_S = (2\pi kT/m_A)^{\frac{1}{2}} \qquad (maximum \; value)}$$

   The assumption of spherical catalyst particles was only a convenience in deriving this formula: the result actually holds for any kind of surface. It has been verified experimentally for the reverse process of desorption by measuring the rate of evaporation of mercury into a vacuum.

   This upper limit rate constant can be modified to give a rate constant for adsorption which takes into account the idea that adsorption may not occur every time that a molecule arrives at the surface. We write the probability of adsorption (called the *sticking coefficient*) in the form $p \exp(-\varepsilon/kT)$ by analogy to the probability of reaction which arose in the collision theory of homogeneous gas reactions. Then the rate of adsorption can be written

$$R_{ads} = k_{ads} \mathscr{A} \, ^1N_A$$

where

$$\boxed{k_{ads} = (2\pi kT/m_A)^{\frac{1}{2}} p \exp(-\varepsilon/kT)}$$

As in the collision theory of homogeneous gas reactions, p expresses the idea that the arriving molecule must arrive at a favorable location of the surface in order to be adsorbed, and the Boltzmann factor $\exp(-\varepsilon/kT)$ indicates that there may be an energy barrier to adsorption.

EXERCISES

7-4 Another illustration of the *principle of detailed balancing* (Exercise 3-19) is given by the detailed mechanism of adsorption. Assume that there are two types of adsorbed molecules, A_{pa}, which are loosely bound to the surface (*physical adsorption*), and A_{ca}, which are strongly bound (*chemical adsorption*). Show that the mechanism

$$A \rightleftarrows A_{pa}$$

$$A_{pa} \rightarrow A_{ca}$$

$$A_{ca} \rightarrow A$$

will not give a correct equilibrium constant. (It has been found experimentally that physical adsorption usually precedes chemical adsorption. By the principle of detailed balancing, desorption must also occur through a physically adsorbed intermediate; that is, $A_{ca} \rightarrow A$ in the above mechanism must be replaced by $A_{ca} \rightarrow A_{pa}$.)

7-5 (a) Use the maximum adsorption rate formula $^1R_E = (2\pi kT/m_A)^{\frac{1}{2}} \mathscr{A}\ ^1N_A$ to calculate the minimum time required to form a complete monolayer of water on a surface if $p_{H2O} = 10^{-6}$ torr. Assume that each water molecule occupies $10\ \text{Å}^2$ and that all arriving molecules join the monolayer. (Actually a sticking coefficient of 0.03 was measured for water molecules arriving at a water surface.) (b) What is the average lifetime of a water molecule on the surface of water in equilibrium with its vapor at 298°? The vapor pressure of water at 298°K is 23.8 torr (cf. Exercise 9-12).

7-6 Consider the heterogeneous (gas–solid) catalysis mechanism $A \rightarrow A_{ads} \rightarrow B_{ads} \rightarrow B$; let the heats of adsorption (i.e., ΔH° for reactions $A \rightarrow A_{ads}$ and $B \rightarrow B_{ads}$ be E_A and E_B, the heat of the surface reaction be ΔH_{AB}, and the activation energies of the three steps be E_1, E_2, and E_3. (a) Draw a graph of energy versus reaction coordinate (similar to Figure 3-1) indicating each of the energetic quantities. (b) What activation energy will be measured if (1) step 1, (2) step 2, (3) step 3 is rate limiting?

7-7 Adsorption of diatomic molecules on metals may give adsorbed atoms, that is, $A_2 \rightarrow 2A_{ads}$. (a) Show that the Langmuir isotherm for this process is $\theta = bp^{\frac{1}{2}}/(1 + bp^{\frac{1}{2}})$. (b) Write out an expression for b in terms of partition functions for this case.

7-8 Derive a rate law for the heterogeneous mechanism $A \rightarrow 2A_{ads}$, $B \rightarrow B_{ads}$, $A_{ads} + B_{ads} \rightarrow$ products, assuming that A is weakly adsorbed, B is strongly adsorbed, and that the third step is rate limiting.

7-2 ELECTRODE REACTIONS

Electrochemical reactions have played an important role in this history of chemistry, and in the history of physics as well. Indeed, one can justifiably trace the beginnings of the whole subject of physical chemistry to the invention of the voltaic cell by Count Alessandro Volta in 1800. In the nineteenth century, as the technology of electrical measurements improved, chemists used electrochemical experiments for establishing many key concepts of chemistry and physics. Stoichiometry, oxidation and reduction, ionization in solutions, chemical equilibrium, and thermochemistry all received much of their original support from electrochemical discoveries. In addition, electrochemistry contributed heavily to the discovery of elements and in the synthesis of many new compounds. During the twentieth century, as electrical power for industry became economical, large-scale electrochemical processes were developed for producing a wide variety of important industrial chemicals. Electrochemical methods also have many important applications in analytical chemistry.

Chemical kinetics research in the field of electrochemistry is centered on the investigation of mechanisms of electrode reactions. We shall find that there are several similarities between reactions at solution–electrode interfaces and reactions at gas–solid interfaces. The mechanisms of both can be subdivided into steps involving mass transfer (convection or diffusion), adsorption–desorption, and reaction at the interface. However, there are two important differences between the two cases. First, reactions at electrodes proceed under the influence of enormous electrical fields. At a solution–electrode interface the electrical potential may fall by as much as a volt over a distance comparable to atomic dimensions (i.e., about 1 V/Å). This is an electric field of 10^8 V/cm, which obviously can have a great influence on the properties of molecules and ions near the electrode. The second distinction is that most electrode reaction mechanisms include a number of homogeneous solution reactions preceding and/or succeeding the reactions at the interface. The availability of methods for making accurate electrical measurements of reaction rates, as well as for varying the reaction conditions by electrical means, compensate for the additional complications.

To give an idea of the scope of mechanism studies in electrochemistry, let us consider the electrochemical reduction of polyaromatic compounds. Anthracene, for example, is reduced electrochemically in the presence of a proton donor to dihydroanthracene by the reaction

$$C_{14}H_{10} + 2e^- + 2H^+ \rightarrow C_{14}H_{12}$$

where the flow of electrical current is indicated by the electron symbol e^-. The mechanism of this reaction in dimethylformamide solvent, in presence of phenol as a proton donor, is thought to be as follows:

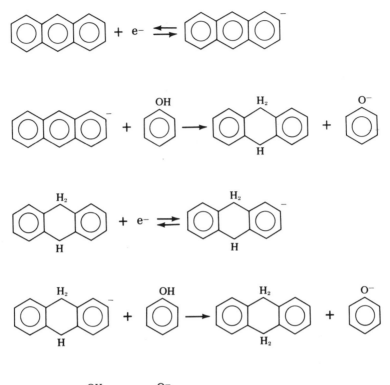

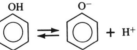

The two electrode reactions forming the anthracene anion and the mono-hydroanthracene radial anion are rapid and reversible. The rate limiting step is the protonation of the anthracene anion by phenol. Since the mono-hydroanthracene radical is more readily reduced electrochemically than anthracene itself, the reduction steps subsequent to the first protonation are rapid. The last elementary reaction represents the equilibrium between phenol and the phenolate anion, which is maintained in experiments by a set of elementary reactions involving the species used to buffer the pH of the solution.

In a mechanism study on this reaction, an electrochemist would investigate the reaction order by studying the effects of concentration changes upon reaction rates. He would also investigate the effects of temperature and solvent changes, the relative reaction rates with different proton donors, and so forth—all the methods one would use in studying homogeneous solution reactions. In addition to these procedures, the electrochemist would study the relationship between the current, time, and voltage of the electrochemical cell in which the reaction is taking place while varying the concentrations of the reactants and the temperature. These methods are only available in electrochemistry, and lend a special power to electrochemical methods of studying mechanisms. In the example we have chosen, the rate limiting step happens to be a homogeneous reaction taking place away from the solution–electrode interface. Electrochemical methods, therefore, provide kinetics data on homogeneous solution reactions as well as on reactions taking place at a solution–electrode interface.

In this section we do not discuss particular electrochemical reaction mechanisms in any detail. Instead we consider the theoretical interpretation of electrode reaction rates. Applications of the theory to interpretations of a variety of electrode reactions can be found in the Supplementary References on electrochemistry at the end of the chapter. These references should also be consulted for descriptions of the application of fast reaction techniques to mechanism studies in electrochemistry.

Elementary reactions at an electrode can lead to current flow in an external electrical circuit only if electrons are added to or taken from the electrode. Let us write such an elementary electrode reaction in the form

$$O + ne^- \rightleftarrows R$$

where O represents some reducible species in the solution, R represents some oxidizable species in the solution, and the stoichiometric coefficient n is the number of electrons transferred to or from the electrode at each occurrence of the elementary reaction. At least one of O and R must be an ion, and frequently both are ions. An example would be

$$Fe^{3+} + e^- \rightleftarrows Fe^{2+}$$

For each mole of electrons transferred when the reaction proceeds in the forward direction, a negative total charge $Q = -n\mathscr{F}$ flows from the external circuit, where $\mathscr{F}$ is the electrical charge of one mole of electrons, 96,487 coulombs. Let the net rate of the general reaction *per unit area $\mathscr{A}$ of electrode surface* be

$$R_{net} = \frac{-dn_0}{dt}$$

where n_0 is the number of moles of O being reduced on one square centimeter of electrode surface.

The current flow in the external circuit is

$$i_{net} = \frac{dQ}{dt}$$

Since $dQ/dt = n\mathscr{F}\mathscr{A}(dn_0/dt)$, the current and the net reaction rate are proportional to one another

$$i_{net} = n\mathscr{F}\mathscr{A}R_{net}$$

In electrochemistry it is convenient to consider the net current as the algebraic sum of currents caused by the forward and reverse reactions, that is, by the reduction and oxidation directions, respectively (cf. Section 2-1). Current resulting from reaction in the reduction direction is called *cathodic current*

$$i_c = n\mathscr{F}\mathscr{A}R_{red}$$

Current resulting from reaction in the oxidation direction is called *anodic current*

$$i_a = n\mathscr{F}\mathscr{A}R_{ox}$$

The net current flow is then

$$i_{net} = i_c - i_a = n\mathscr{F}\mathscr{A}(R_{red} - R_{ox})$$

For most purposes it is appropriate to consider the entire current flow at an electrode to be due to a single elementary reaction that is first order with respect to the oxidizable species O in the reducing direction and first order with respect to the reducible species R in the oxidizing direction

$$R_{red} = k_{red}\,[O]_{x=0}$$

$$R_{ox} = k_{ox}\,[R]_{x=0}$$

The subscripts $x = 0$ indicate that the reaction rates depend on the concentrations right at the solution–electrode interface. These concentrations can be substantially different from the corresponding concentrations in the bulk of the solution. In many experiments the solution is stirred rapidly in order to keep the concentrations at the interface as close as possible to the bulk solution values. Since the rates R_{red} and R_{ox} are expressed in moles cm^{-2} sec^{-1} units, the rate constants k_{red} and k_{ox} are in cm/sec units.

A difference in electrical potential exists between the electrodes of an electrochemical cell whether or not current is flowing. This difference is measured by a *potentiometer* and expressed in units of *volts*. In electrochemistry one distinguishes between the terms *voltage* and *potential* as follows. *Voltage* refers to the actual value of the electrical potential difference between the two electrodes regardless of the conditions in the cell. *Potential*, on the other hand, refers to the actual electrical potential difference minus that part of the electrical potential difference that arises from the fact that a current is flowing through a resistance. The correction term is called the *ohmic* or *iR drop* and calculated from Ohm's law in the form $V = iR$. If no current is flowing the voltage and the potential of a cell are of course the same. We describe later a special experimental arrangement in which a cell voltage is equal to its cell potential even though current flows through one of the electrodes.

We now wish to relate the voltage of an electrochemical cell to the net current flowing, in order to set the stage for introducing the fundamental equations of electrochemical kinetics. First we recall the thermodynamic relationship that describes the equilibrium situation in which no current flows. The voltage of an electrochemical cell that is measured when no current flows, the *reversible cell potential* $\mathscr{E}_{\text{rev}}$, is related to the Gibbs free energy change of the cell reaction ΔG by the equation

$$\Delta G = \Sigma v_i \bar{G}_i = -n\mathscr{F}\mathscr{E}_{\text{rev}}$$

In this equation we have assumed a generalized stoichiometric equation (as in Section 2-1) and denoted the partial molar free energy of the ith reactant by $\bar{G}_i$. The partial molar free energy $\bar{G}_i$ is related to the activity of the ith reactant a_i by

$$\bar{G}_i = \bar{G}^0{}_i + RT \ln a_i$$

where $\bar{G}_i{}^0$ is the standard partial molar free energy. Calculating $\Delta G = \Sigma v_i \bar{G}_i$ from this equation gives

$$\Delta G = \Delta G^0 + RT \ln \prod a_i^{v_i}$$

Substitution of this equation into the relation between free energy change and cell voltage gives the *Nernst equation*

$$\boxed{\; \mathscr{E}_{\text{rev}} = \mathscr{E}_0 - \frac{RT}{n\mathscr{F}} \ln \prod a_i^{v_i} \;}$$

When current flows the cell voltage does not equal $\mathscr{E}_{\text{rev}}$. If the cell is operated as a battery, with current drawn *from* it, then the voltage is *less than* $\mathscr{E}_{\text{rev}}$; if current is forced through the cell in the direction opposite to the spontaneous flow that operation as a battery produces, the voltage is *greater than* $\mathscr{E}_{\text{rev}}$. The difference between the voltage and $\mathscr{E}_{\text{rev}}$ is called the *cell overvoltage*.

In a typical battery, for instance a flashlight battery or an automobile storage battery, the difference between the voltage and $\mathscr{E}_{rev}$ is due to kinetic processes occurring at both of the two electrodes that make up the cell, as well as to mass transfer effects and iR drop. This would be an awkward situation for scientific research since there would be no way to separate effects at the two electrodes by electrical measurements. This difficulty can be circumvented by a special arrangement, one example of which is in Figure 7-1, that permits the kinetics to be studied at a single electrode. The

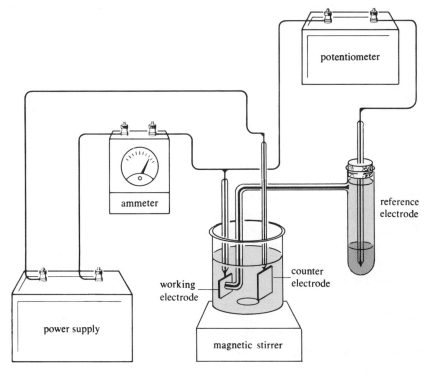

FIGURE 7-1 *A typical experimental apparatus used to study electrode reaction kinetics. The nature of the working electrode and the solution in the beaker are, of course, determined by the electrode reaction one wishes to study. For studying the kinetics of the reaction* $Zn^{2+} + 2e^- \rightleftarrows Zn$, *for instance, the working electrode would be made of zinc and the solution would contain a known concentration of zinc ions. The counter electrode can be made of any metal that does not react with the solution. The capillary bridge connecting the reference electrode to the working electrode is placed extremely close to the surface of the working electrode in order that it will not be influenced by the* iR *drop across the beaker. For measuring polarization curves the power supply, ammeter, and potentiometer can be simple electrical devices. If fast reaction methods are to be employed, however, these parts of the circuit must be sophisticated instruments capable of fast control and measurement.*

current through the electrode under study, called the *working electrode*, passes almost entirely through a large *counter electrode*.   A third electrode, called the *reference electrode*, is added in such a way that very little current flows through it.   The voltage measured between the working electrode and the reference electrode is called the *electrode potential* of the working electrode with respect to that particular reference electrode.   If no current flows through the working electrode, and equilibrium is established, then the voltage measured is the *reversible electrode potential*.   It can be converted to the familiar *half-cell potential* of the working electrode by adding to it the potential of the reference electrode measured with respect to the standard hydrogen electrode.

The cell overpotential $\mathscr{E} - \mathscr{E}_{rev}$ that is measured between the working electrode and the reference electrode when current flows between the working electrode and the counter electrode is called the *electrode overpotential*, or simply *overpotential*, and denoted η.   Since the reference electrode is unaffected by the current flow through the working electrode, *the overpotential characterizes the behavior at the working electrode alone*.   By measuring the dependence of overpotential on the current flowing through the working electrode, electrochemists are able to draw conclusions about the kinetics of the reactions occurring at the working electrode.   To show how this is done we have to develop a theory of electrode reaction rates.

Theory of electrode reaction rates.   The overpotential η directly represents a driving force keeping the reaction(s) at the working electrode out of equilibrium.   It was discovered by J. Tafel in 1905, and has been confirmed many times since, that a relationship of the form

$$\eta = a + b \log i$$

provides a close fit to a wide variety of experimental data on the dependence of overpotential on current.   The Tafel equation can be rearranged to show the current as a dependent variable

$$i = a' e^{-\eta/b'}$$

Since i is a measure of the reaction rate, the form of this equation is reminiscent of the Arrhenius rate constant formula.   It suggests that η is a measure of the effect of the electrical potential in overcoming a barrier to reaction, in analogy, for example, to the role of relative translational energy in overcoming the barrier to reaction in a bimolecular gas reaction.   Our task now is to formulate a theoretical interpretation in which the relationship between η and i is expressed in terms of physical quantities; that is, we must identify the empirical parameters a and b with physical quantities.

This task is most easily accomplished by utilizing the thermodynamic formulation of activated complex theory (Section 6-1c). Let us continue to consider the elementary electrochemical reaction $O + ne^- \rightleftarrows R$, having a forward rate constant k_{red} and a reverse rate constant k_{ox}. In terms of activated complex theory these rate constants depend on free energies of activation $\Delta G_{red}^{0\dagger}$ and $\Delta G_{ox}^{\dagger 0}$ according to

$$k_{red} = \frac{kT}{h} \exp(-\Delta G_{red}^{0\dagger}/RT)$$

$$k_{ox} = \frac{kT}{h} \exp(-\Delta G_{ox}^{0\dagger}/RT)$$

When the overpotential η is zero, the forward and reverse rates are equal

$$R_{red} = R_{ox}$$

$$k_{red} [O]_{x=0} = k_{ox} [R]_{x=0}$$

$$\frac{kT}{h} \exp(-\Delta G_{red}^{0\dagger}/RT) [O]_{x=0} = \frac{kT}{h} \exp(-\Delta G_{ox}^{0\dagger}/RT) [R]_{x=0}$$

Converting to logarithmic form and rearranging yields for the equilibrium condition

$$RT \ln \left(\frac{[R]_{x=0}}{[O]_{x=0}} \right)_{eq} = (\Delta G_{ox}^{0\dagger} - \Delta G_{red}^{0\dagger})_{eq}$$

A second formula for the left-hand side of this equation can be obtained from the thermodynamic equations for electrochemical cells. The partial molar free energy difference of R and O is given by

$$\Delta G = \bar{G}_R - \bar{G}_O$$
$$= \bar{G}_R{}^0 - \bar{G}_O{}^0 + RT \ln (a_R/a_O)$$

In electrochemistry one frequently ignores the distinction between activity and concentration. Accepting this ignoration and rearranging yields

$$RT \ln \frac{[R]}{[O]} = \Delta G - (\bar{G}_R{}^0 - \bar{G}_O{}^0)$$

If we further assume that the concentrations at the electrode are equal to the bulk solution values, then

$$RT \ln \left(\frac{[R]_{x=0}}{[O]_{x=0}} \right)_{eq} = \Delta G_{eq}(\text{working electrode}) - (\bar{G}_R{}^0 - \bar{G}_O{}^0)$$

where the eq subscript and the (working electrode) notation have been added to emphasize that this equation was derived on a thermodynamic basis, thus assuming that there is equilibrium in the system and, in particular, at the working electrode. If we set the kinetic and thermodynamic expressions for $RT \ln ([R]_{x=0}/[O]_{x=0})_{eq}$ equal to one another, we obtain after trivial rearrangement

$$\Delta G_{eq}(\text{working electrode}) = (\Delta G_{ox}^{0\dagger} - \Delta G_{red}^{0\dagger})_{eq} + (\bar{G}_R{}^0 - \bar{G}_O{}^0)$$

The quantity ΔG_{eq} (working electrode) given here is only one part of the free energy change of the cell reaction in the cell formed by the working electrode and the reference electrode; the remaining parts are obviously given by similar expressions involving the free energies of the species participating in the electrode reaction at the reference electrode and in any other homogeneous reactions that may be part of the mechanism of the reference cell reaction. Symbolically, ΔG_{eq} (reference cell) = ΔG_{eq} (working electrode) + ΔG_{eq} (rest of reference cell). When the reversible potential of the reference cell changes (due to current flow) from $\mathscr{E}_{rev}$ to $\mathscr{E}_{rev} + \eta$, the equilibrium relationship

$$\Delta G \text{ (reference cell)}_{eq} = -\eta \mathscr{F} \mathscr{E}_{rev}$$

is *assumed* to retain its form and read

$$\Delta G(\text{reference cell})_{with\ current\ flow} = -n\mathscr{F}(\mathscr{E}_{rev} + \eta)$$

or

$$\delta[\Delta G \text{ (reference cell)}] = -n\mathscr{F}\eta$$

where the δ symbol denotes the change that occurs when current flows and overpotential appears. Having arranged our electrochemical cell such that the current flowing through the reference electrode is very small, all of the change in ΔG (reference cell) when current flows through the working electrode appears in ΔG (working electrode) itself. Thus

$$\delta[\Delta G \text{ (reference cell)}] = \delta \, [\Delta G \text{ (working electrode)}]$$

Referring to the equation for ΔG_{eq} (working electrode) preceding this paragraph, and noting that $(\bar{G}_R{}^0 - \bar{G}_O{}^0)$ is unaffected by current flow, we conclude that

$$\delta[\Delta G(\text{working electrode})] = \delta[\Delta G_{ox}^{0\dagger} - \Delta G_{red}^{0\dagger}]$$

The effect of current flow is to change the difference *between the free energies of activation for the forward and reverse electrode reaction.* Substituting $-n\mathscr{F}\eta$ for the left-hand side of this equation yields

$$\boxed{n\mathscr{F}\eta = \delta[\Delta G_{red}^{0\dagger} - \Delta G_{ox}^{0\dagger}]}$$

This equation represents the fundamental interconnection between overpotential and electrode reaction rates. For reasons that will become apparent later, it is customary to consider the overpotential η to act separately upon $\Delta G_{red}^{0\dagger}$ and $\Delta G_{ox}^{0\dagger}$. In conventional notation, a fraction α of the overpotential acts to decrease $\Delta G_{red}^{0\dagger}$, while the remaining $(1 - \alpha)$ of the overpotential acts to increase $\Delta G_{ox}^{0\dagger}$. (Note the sign convention here: the forward reaction of reduction is favored by negative overpotential, which increases the cathodic current i_c while decreasing the anodic current i_a. Positive values of η correspond to favoring the reverse reaction of oxidation. The convention on α is unchanged, but since η is positive, $G_{red}^{0\dagger}$ is then increased while $G_{ox}^{0\dagger}$ is decreased.) Summarizing our equations we have

$$\boxed{\begin{aligned} \alpha n\mathscr{F}\eta &= \delta(\Delta G_{red}^{0\dagger}) \\ (\alpha - 1)n\mathscr{F}\eta &= \delta(\Delta G_{ox}^{0\dagger}) \end{aligned}}$$

We are now in a position to determine the relationship between current and overpotential in the ACT formalism. Let the equilibrium (no current) value of $\Delta G_{red}^{0\dagger}$ be $(\Delta G_{red}^{0\dagger})_{eq}$; when current flows $\Delta G_{red}^{0\dagger} = (\Delta G_{red}^{0\dagger})_{eq} + \alpha n\mathscr{F}\eta$. The rate constant for the reducing reaction is

$$k_{red} = \frac{kT}{h} \exp\{-[(\Delta G_{red}^{0\dagger})_{eq} + \alpha n\mathscr{F}\eta]/RT\}$$

$$= k_{red}^0 \exp(-\alpha n\mathscr{F}\eta/RT)$$

where k_{red}^0 is the rate constant at equilibrium. For the oxidizing reaction we obtain

$$k_{ox} = \frac{kT}{h} \exp\{-[(\Delta G_{ox}^{0\dagger})_{eq} + (\alpha - 1)n\mathscr{F}\eta]/RT\}$$

$$= k_{ox}^0 \exp[(1 - \alpha)n\mathscr{F}\eta/RT]$$

With the rate constants in hand it becomes a simple matter to obtain a current-overpotential equation.

$$
\begin{aligned}
i_{net} &= n\mathscr{F}\mathscr{A}(R_{red} - R_{ox}) \\
&= n\mathscr{F}\mathscr{A}\{k^0_{red}[O]_{x=0}\exp(-\alpha n\mathscr{F}\eta/RT) \\
&\quad - k^0_{ox}[R]_{x=0}\exp[(1-\alpha)n\mathscr{F}\eta/RT]\}
\end{aligned}
$$

This is the fundamental equation of electrochemical kinetics. Let us note two of its important features. First, if η is zero, i_{net} must also be zero. This implies that

$$k^0_{red}[O]_{x=0} = k^0_{ox}[R]_{x=0}$$

or

$$\frac{k^0_{red}}{k^0_{ox}} = \frac{[R]_{x=0}}{[O]_{x=0}}$$

which we recognize as our familiar statement that the ratio of forward and reverse rate constants equals an equilibrium constant. Second, we note that the arguments of the two exponential functions have opposite signs. This means that one of the two exponentials will become substantially smaller than the other one whenever the overpotential is substantially different from zero. Suppose that η is sufficiently positive that the oxidizing current term can be neglected compared with the reducing current. Then

$$
\begin{aligned}
i_{net} = i_c &= n\mathscr{F}\mathscr{A}R_{red} \\
&= n\mathscr{F}\mathscr{A}k^0_{red}[O]_{x=0}\exp(-\alpha n\mathscr{F}\eta/RT)
\end{aligned}
$$

Inspection of the form of this equation shows that we have derived the Tafel equation $i = a'\exp(-\eta/b')$, which was given as an empirical finding at the start of this section, and identified the empirical constants as

$$
\begin{aligned}
a' &= n\mathscr{F}\mathscr{A}k^0_{red}[O]_{x=0} \\
b' &= RT/\alpha n\mathscr{F}
\end{aligned}
$$

The constant a', which is the hypothetical cathodic current flowing when $\eta = 0$, is called the *exchange current*. It provides the value of k^0_{red} if n and

$[O]_{x=0}$ are known. The constant b' tells the fraction of the overpotential that affects the cathodic barrier to reaction (if n is known). Our theory has thus given us a framework for interpreting the parameters of the Tafel equation.

Three aspects of our derivation deserve further comment. One has to do with replacing activities with concentrations. This is only true as an approximation, and we found in studying the primary salt effect (Section 6-2) that closer inspection reveals that activity coefficients influence rates. The primary salt effect is just as important in electrode kinetics as it is in homogeneous solution kinetics. The rate constants k^0_{red} and k^0_{ox} depend on ionic strength in a manner that is analogous to the homogeneous solution reaction case discussed in Section 6-2.

The second aspect of our derivation that deserves further attention is our assumption that the electrode reaction involves two species present in the solution. Some of the most important electrode reactions are *corrosion* reactions, in which the reduced species is the metal electrode itself. For example

$$Zn^{2+} + 2e^- \rightarrow Zn$$

The metal can also react electrochemically with a species in solution, or with a product of the electrochemical reaction. Electrolysis of brine with a mercury cathode, for example, leads to

$$Na^+ + e^- \rightarrow Na \, (Hg \text{ amalgam})$$

For these situations the first order rate $k_{ox}[R]_{x=0}$ becomes a zero order rate k_{ox}; the procedure of derivation is otherwise unaffected.

The third aspect of our derivation in need of comment is our assumption that the process $O + ne^- \rightarrow R$ was an elementary reaction. Although our method of derivation required this assumption in order that the ACT equation could be applied, the equations derived are actually valid even if the electrode reaction involves a mechanism. If a mechanism is involved, however, the parameters α and k^0 refer in a complex way to the interconnection between overpotential and the several elementary reactions concerned.

Electrode reaction kinetics. The starting points for a rate study on an electrode reaction are identification of the reaction products and a graph of η versus i_{net}. Such graphs are called *polarization curves* by electrochemists. They serve as convenient summaries of the macroscopic electrochemistry of electrode reactions. Converted to a semilogarithmic form, the polarization curve also provides the parameters of the Tafel equation. An example is given in Figure 7-2.

Kinetics experiments on electrode reactions can be carried out using any of the three variables—concentration, current, or overpotential—as the independent ones. A complete kinetics study would naturally include all three possibilities. Included under the concentration variable category would be investigation of the effects of pH and ionic strength.

A major simplification of data analysis in electrochemistry is afforded by the equation $i_{net} = n\mathscr{F}\mathscr{A}R_{net}$, that is, by the fact that a current measurement is fully equivalent to a net rate measurement. Determining the influence of

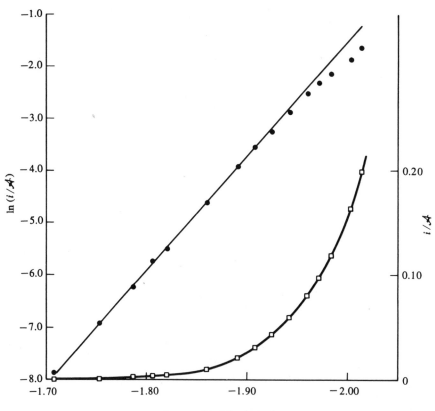

$\mathscr{E}$ versus Hg SO$_4$ electrode

FIGURE 7-2   *An example of a polarization curve and a Tafel plot. The electrode reaction is* $2H^+ + 2e^- \rightarrow H_2(g)$ *on a mercury surface in* $2N\ H_2SO_4$, *and the data points were obtained by Tafel himself (cf. Supplementary References). The current density is expressed in* A cm^{-2} *and the electrode potential is expressed in volts versus a mercuric sulfate reference electrode. Note the obvious departure from linearity in the Tafel plot as mass transfer effects appear at high values of* ln(i/𝒜). *In this case no departure from linearity owing to exchange current appears at low values of* ln(i/𝒜).

concentration changes on i at constant η in the exact same procedure that we have dealt with before in determining reaction order. In electrochemistry one merely has the great simplification that R_{net} can be read from an ammeter. Reaction orders for all reactants whose concentration can be varied are, therefore, obtainable from logarithmic graphs of current versus concentration. Normally, the only concentration that cannot be varied is the solvent concentration. Quite frequently, however, acid–base or complex ion equilibria in the solution can enter to obscure the connection between order determination and the mechanism of the electrode reaction.

It is unfortunately the case in electrochemistry, as in the rest of reaction kinetics, that the intrinsically most interesting electrode reactions prove to be the ones whose mechanisms are most difficult to unravel. The electrochemical corrosion of iron, for example, represented in the simplest fashion as $Fe \rightarrow Fe^{2+} + 2e^-$, proves to be extremely sensitive to pH and to concentrations of ions present in the solution. An even more elusive mechanistic problem in electrochemistry is the *hydrogen overvoltage problem*, which concerns the mechanism of hydrogen evolution on metal electrodes. The reversible electrode potential of a hydrogen electrode can only be measured by extrapolation to $i = 0$; if a measureable current passes through the hydrogen electrode, η exceeds zero by large amounts, which depend strongly on the metal used to make the electrode. If 10^{-2} A flow through an electrode of 1 cm^2 area, and $[H^+] = 1.0$ mole/liter, $\eta = 0.035$ V for "platinized" platinum, 0.56 V for iron, 0.76 V for silver, and 1.10 V for mercury. Despite decades of intensive research on this reaction, the identities of the rate limiting step(s) in the mechanism, which is different for different electrodes, are quite uncertain. Some suggestions that have been advanced for the rate limiting step are

$$H_2O + e^- \rightarrow e^-_{aq}$$

$$H_{ads} + H_{ads} \rightarrow H_{2ads}$$

$$H_2O_{ads} + H_2O_{aq} \rightarrow H_3O^+_{ads} + OH^-_{aq}$$

$$H_3O^+_{aq} + e^- \rightarrow H_{ads} + H_2O_{ads}$$

$$H_3O^+_{aq} + H_{ads} + e^- \rightarrow H_2(gas) + H_2O_{aq}$$

Examples of electrochemical data taken on simpler systems are introduced in the exercises at the end of this section.

Electronics technology provides the experimenter on electrode reactions with a variety of specialized methods of investigating the relationship between η and i at constant $[O]$ and $[R]$. Many of these methods fall into the fast reaction category introduced in Chapter 9.   Unfortunately, the diversity of experimental techniques in electrode kinetics, and the indirectness of many of the measurements made, force us to refrain from describing them in much detail.   The response of i to rapid changes in η, or vice versa, is the simplest of the fast reaction methods used in electrochemistry.   In the case of step changes in η or i, the techniques are called *potentiostatic* and *galvanostatic*, respectively.   For extremely rapid reactions, a short pulse of current can change the charge of the capacitor formed by the electrode surface and the layer of ions next to it; the subsequent variation of η provides information on fast steps at the surface.   This technique is called a *coulostatic* one.   The *polarographic* techniques used in electroanalytical chemistry can also be applied for kinetics experiments.

Mass transfer effects.   In our development of the theory of electrode reaction rates we tacitly assumed that transport of O and R to and from the solution–electrode interface did not constitute a rate limiting step.   The net current cannot continue indefinitely to increase exponentially with increasing overvoltage, however: sooner or later a point is reached where the current begins to be limited by the rate at which reactants reach the electrode or by the rate at which products leave the electrode.   This effect is known as *concentration polarization*.   An example of it appears in the failure of the Tafel plot in Figure 7-2 to agree with the experimental data at the high current densities.

There are three processes by which matter can be supplied to or removed from an electrode.   Transport by *diffusion* would be the solution analog to the maximum adsorption rate problem at a gas–solid interface discussed in Section 7-1c.   *Migration* refers to the motion of ions under the influence of an electric field.   *Convection* refers to mass transport to the electrode by motion of the bulk solution, for example by stirring.   A theoretical discussion of mass transfer including the details of each of these processes would be a complex undertaking.   We develop here a simplified treatment that describes the essential influence of mass transfer on electrode reaction rates without regard to which of the three processes are responsible for it in a particular case.   The books cited as Supplementary References on electrochemistry have extensive discussions of the details.

Let us suppose that the rate of a reduction reaction is being measured at an overpotential sufficiently high that the rate of mass transfer of O to the electrode is a major limiting factor.   The rate of mass transfer to the electrode surface will be proportional to the concentration gradient at the electrode.   We assume that this gradient is approximately proportional to the difference $[O] - [O]_{x=0}$.   (This proportionality holds exactly for some models of transport processes.)   Calling the constant of proportionality m_O we have

$$R_{m.t.} = m_O([O] - [O]_{x=0})$$

The cathodic current then depends in the same way on $([O] - [O]_{x=0})$

$$i_c = n\mathscr{F}\mathscr{A}R_{\text{m.t.}} = n\mathscr{F}\mathscr{A}m_0([O] - [O]_{x=0})$$

The maximum current, which is called the *limiting current* and denoted i_ℓ, is obtained when $[O]_{x=0}$ approaches zero

$$i_\ell = n\mathscr{F}\mathscr{A}m_0[O]$$

The ratio i_c/i_ℓ is given by

$$\frac{i_c}{i_\ell} = \frac{n\mathscr{F}\mathscr{A}m_0([O] - [O]_{x=0})}{n\mathscr{F}\mathscr{A}m_0[O]} = 1 - \frac{[O]_{x=0}}{[O]}$$

Solving this equation for $[O]_{x=0}$ we obtain

$$\boxed{[O]_{x=0} = [O](1 - i_c/i_\ell)}$$

This equation describes in a simple way the essential effect of mass transfer in limiting the rate of the electrode reaction. At low values of cathodic current, i_c/i_ℓ is small and $[O]_{x=0} = [O]$. At high values of cathodic current, i_c/i_ℓ approaches unity and $[O]_{x=0}$ approaches zero. The effect on the cathodic current can be found by substituting the equation for $[O]_{x=0}$ into the general equation for i_c, giving

$$\boxed{i_c = n\mathscr{F}\mathscr{A}k_{\text{red}}^0[O](1 - i_c/i_\ell)e^{-\alpha n\mathscr{F}\eta/RT}}$$

When i_c/i_ℓ is small, this equation is our original equation for the cathodic current except that $[O]_{x=0}$ has been replaced by $[O]$. If the cathodic current is increased to the point that it begins to be comparable to the limiting current, the exponential relationship between current and over-potential begins to fail. To obtain an explicit equation for the cathodic current we rearrange the above equation to

$$\frac{1}{i_c} = \frac{e^{\alpha n\mathscr{F}\eta/RT}}{n\mathscr{F}\mathscr{A}k_{\text{red}}^0[O]} + \frac{1}{i_\ell}$$

which is useful for plotting data if not for providing a transparent formula. As an equation for η we have after trivial rearrangement

$$\eta = \frac{RT}{\alpha n\mathscr{F}}\left[\ln(n\mathscr{F}\mathscr{A}k_{\text{red}}^0[O]) + \ln\left(\frac{1}{i_c} - \frac{1}{i_\ell}\right)\right]$$

Analogous expressions for the anodic current can be derived by the same procedure. The results are

$$[R]_{x=0} = [R](1 - i_a/i_\ell)$$

$$i_a = n\mathscr{F}\mathscr{A}k_{ox}^0 [R](1 - i_a/i_\ell) e^{(1-\alpha)n\mathscr{F}\eta/RT}$$

In these two equations the limiting current i_ℓ refers to the *anodic* limiting current. Since mass transfer effects appear for cathodic or anodic currents that are much larger than the exchange current (for all but the fastest electrode reactions), it is not necessary to combine our equations to find an expression for the net current.

EXERCISES

7-9   Consider the general electrode reaction $O + ne^- \leftrightarrows R$ with $n = 1$, $[O] = [R] = 10^{-3}$ molar, $k_{red}^0 = 10^{-4}$ cm/sec, and $\alpha = 0.5$.   (a) Calculate the exchange current density.   (b) Construct a polarization curve over the current density range -6×10^{-4} to 6×10^{-4} A cm^{-2}.   (c) Draw a semilogarithmic (Tafel) graph over the same range.

7-10   For the electrode reaction $2H^+ + 2e^- \rightarrow H_2(g)$ on a mercury electrode in dilute H_2SO_4, η was found to be 0.72 V at a current density of $i/\mathscr{A} = 10^{-4}$ A cm^{-2}, while $\eta = 0.54$ V at $i/\mathscr{A} = 10^{-6}$ A cm^{-2}.   Calculate the parameters of the Tafel equation, and evaluate α.

7-11   At an overpotential of -1.4 V measured against the standard hydrogen electrode the current density for H_2 evolution on a mercury electrode is $i/\mathscr{A} = 6 \times 10^{-2}$ A cm^{-2} at pH $= 2$ and 4×10^{-6} A cm^{-2} at pH $= 6$.   Calculate the order of reaction with respect to H^+.

7-12   The exchange current for copper deposition from a 0.1 molar $CU(NO_3)_2$ solution at 300°K is 10^{-11} A for a 1-cm^2 electrode, and α is 0.55.   Calculate k_{red}^0.

7-13   At constant overpotential of 1.4 V against the normal hydrogen electrode, a platinum electrode in 7.5-molar H_2SO_4 solutions of Mn^{3+} and Mn^{4+} ions passes the following currents:

$[Mn^{3+}]$	$[Mn^{4+}]$	i_{net}
0.01	0.01	1.5×10^{-4}
0.01	0.001	10^{-4}
0.0025	0.01	3.5×10^{-5}

The direction of current flow is such that the cell reaction is $Mn_{aq}^{4+} + 1/2H_2(g) \rightarrow Mn_{aq}^{3+} + H_{aq}^+$.   Find the reaction orders with respect to Mn^{3+} and Mn^{4+}, and propose a mechanism for the electrode reaction $Mn^{4+} + e^- \rightarrow Mn^{3+}$.

7-14   Calculate a, b, α, and k_{red}^0 for $2H^+ + 2e^- \rightarrow H_2(g)$ on a mercury electrode from the data shown in Figure 7-2.   What is the approximate value of the limiting current density?

7-15   Data for the temperature dependence of the kinetic parameters for generation of hydrogen on a mercury cathode are given in the article by J. Tafel cited in the
Supplementary References.   Calculate $\Delta G_{red}^{0\ddagger}$ for $2H^+ + 2e^- \rightarrow H_2(g)$ on mercury
from Tafel's data.

7-3   CATALYSIS BY ENZYMES

Chemical reactions in living things are catalyzed by *enzymes*, which are
protein molecules about 30 to 1000 Å in diameter.   As their size indicates,
their behavior as catalysts is intermediate between the homogeneous and
heterogeneous cases we have considered so far.   The subject of biochemical
kinetics is more or less evenly divided between the study of biochemical
mechanisms and the study of the nature of the enzyme catalysis that governs
the mechanisms.

Enzymes are classified according to the reactions that they catalyze.   Thus
esterases catalyze the hydrolysis of esters, lipases act on fats, arginase catalyzes the hydrolysis of arginine into ornithine and urea.   Close consideration
of enzyme classifications reveals that enzymes may show great *specificity*.
Some of them catalyze only a single reaction.   Urease, for example, rapidly
converts urea into ammonia and carbon dioxide but has never been found
to catalyze any other reaction.   Others may catalyze a group of related
reactions, as the β-galactosidases; still others may show low specificity, as
the lipases, which hydrolyze all sorts of fats.   Quite generally, enzymes act
on optically active molecules and are effective only on one of the optical
isomers.   The specificity of enzymes can far exceed that of the best man-
made catalysts.   One could, in fact, regard enzyme specificity as the essential
characteristic of living organisms, for only a minute fraction of the possible
reaction paths of the molecules found in them will provide the synthetic
and degradative routes necessary to maintain life.

A second noteworthy characteristic of enzymes is their catalytic *speed*.   It
is sometimes difficult to measure enzyme reaction rates on a per-enzyme-
molecule basis because of the difficulties of avoiding poisoning (disablement)
of the active sites or degradation of the proteins, not to mention the equally
or perhaps more difficult problem of obtaining pure enzymes in the first
place.   Nonetheless there are now many reliable rate constants for enzyme-
catalyzed reactions, and most of them are very large compared to solution
reaction rate constants for similar reactions.   The rates of some enzyme
reactions are diffusion controlled.

A final important aspect of enzyme kinetics is great mechanistic *complexity*.
Partial evidence for this was obtained in classical rate measurements which
showed that enzyme reactions have complicated rate laws, high sensitivity
to changes in pH and ionic strength, and large temperature dependence.
More importantly, however, the application of fast reaction methods to

enzyme kinetics has shown that even those reactions which have simple steady state rate laws actually have complex mechanisms.

The subject of enzyme kinetics is conveniently divided into three parts. In *steady state enzyme kinetics* one considers more or less classical experiments in which the over-all rate laws are determined and the products and reactants are carefully identified. A distinguishing characteristic of steady state enzyme kinetics experiments is that the enzyme concentration is low (typically 10^{-7} to 10^{-10} molar) and that the steady state approximation can be applied to the concentration of the enzyme and all species formed when reactants or products are bound to enzymes. In *fast reaction enzyme kinetics* the enzyme concentrations are much higher, and it is possible to investigate the rates of the reactions responsible for establishing the steady state concentrations, as well as any other fast steps. With *x-ray crystallography* it is possible to determine the molecular structure of enzymes and to deduce from the structure not only the location of the catalytically active part (*active site*) of the enzyme molecule, but many of the salient characteristics of the catalytic process as well. The laboratory activities of chemists engaged in each of these three areas of research are almost totally different. Moreover, the information derived from experiments in the three different areas is essentially independent of knowledge derived from experiments in the other two. The true power of research in the field of enzyme kinetics, however, only comes to light when one combines knowledge from all three. The picture that has emerged in recent years for several enzyme reactions leads one to believe that mutually supporting evidence from these independent lines of inquiry will allow greater understanding of the mechanism of enzyme catalysis than of any other area of chemical kinetics. The implications for biological and medical research are obvious.

Our intention here, as in the case of electrode reactions, is to display the essential features of kinetics research in the field rather than to attempt a summary of knowledge that has been gained. It will be expedient to single out as examples a few enzyme reactions that proved to be exceptionally amenable to particular lines of enzyme kinetics research. The examples discussed are therefore not at all typical of the field. The general status of knowledge on more typical enzyme reactions is far less clear than in our examples. It is perhaps only a hope at the present time that continued research will lead to equally enlightening results on a wide variety of enzyme reactions.

Steady state enzyme kinetics. Research on the mechanisms of enzyme reactions has a long history, reaching back almost as far as the first kinetics experiments. Throughout all of the early history of this research, however, enzyme chemists were plagued by difficulties in obtaining pure starting materials. Their enzymes were usually contaminated by other enzymes and catalytically inert substances. Not until 1930 was it possible to obtain the first pure crystals of an enzyme, which happened to be the proteolytic enzyme pepsin. It was 1967 before a complete structural determination of an enzyme, which happened to be a polysaccharide-splitting enzyme called lysozyme, was available, and the announcement of the first total synthesis of an enzyme (ribonuclease) did not appear until 1969. The problem of obtaining pure enzyme preparations and of preventing enzymes from being degraded while standing and during kinetics experiments was a serious deterrent to obtaining accurate rate data. Much of the experimental evidence on which the early theories of enzyme mechanisms were based is therefore of rather uncertain quality in comparison with enzyme kinetics experiments done today. The general form of the mechanistic intepretation of enzyme reactions has, nonetheless, remained much as it was formulated by the early researchers in the field.

The basic mechanism proposed for enzyme reactions, known as the Michaelis–Menten mechanism, is useful for discussing the principal observations and also for providing biochemists with a uniform jargon for reporting experimental data, even though it now appears unlikely that any actual enzyme reaction follows this mechanism. The reactants of enzyme reactions, called *substrates* and denoted by S, are presumed to combine in a reversible biomolecular step with an enzyme E to form an *enzyme–substrate complex* ES. The complex decomposes unimolecularly to yield the product P and regenerate the free enzyme E.

$$E + S \underset{k_{-1}}{\overset{k_1}{\rightleftharpoons}} ES \overset{k_2}{\rightarrow} E + P$$

Our usual steady state approximation must be modified to take into account the circumstance that in steady state enzyme kinetics experiments the concentrations of ES and E may be comparable to one another. If $[E]_0$ is the concentration of enzymatic sites before any substrate is added, then $[E] + [ES] = [E]_0$ at all times. Solution for the steady state rate yields

$$R = \frac{k_2\,[E]_0[S]}{K_m + [S]} = \frac{V\,[S]}{K_m + [S]}$$

At high substrate concentrations the enzyme sites are saturated, step 2 is rate determining, and the reaction is zero order in [S]; at low substrate concentrations step 1 is rate determining, and the reaction is first order in [S]. K_m, which gives the substrate concentration at which R has one-half its saturation value V, is called the *Michaelis constant*. The first order rate constant k_2 is called the *turnover number*.

EXERCISE 7-16   (a) Complete all steps of the derivation in the preceding paragraph.   Relate K_m to the rate constants k_1, k_{-1}, and k_2.   (b) Sketch graphs of $[S]/R$ versus $[S]$, and R^{-1} versus $[S]^{-1}$.   Show how each of your graphs could be used to find V and K_m.   The third graph is known to enzyme chemists as a *Lineweaver–Burk plot*.

The essential characteristic of saturation in steady state enzyme reaction rates is illustrated in Figure 7-3.   At low substrate concentrations the rate is proportional to the substrate concentration, whereas at high substrate concentrations the rate is constant.   There is of course wide variation in the values of the rate constants among enzyme reactions.   The first order rate constant k_2, for example, ranges from 10^{-3} sec^{-1} for pepsin, which acts on peptide bonds, to 10^6 sec^{-1} for catalase, which decomposes hydrogen peroxide.

There are many elaborations of the Michaelis–Menten mechanism, and general methods for deriving the corresponding kinetic equations have been developed.   Two relatively simple extensions are the subjects of Exercises 7-17 and 7-18.   Rate constants have been measured, with varying reliability, for a large number of reactions.   Relative rate constants for enzymes that catalyze a variety of different reactions have been measured with good accuracy.   Out of such studies emerge a number of interesting conclusions. The first is the previously mentioned fact that there is a wide variation in enzyme reaction rates.   The second is that small changes in temperature or pH can have a large effect on rate constants even though there is no indication of any mechanism change.   In the α-chymotrypsin-methyl hydrocinnamate system, for example, catalytic activity appears only in the pH

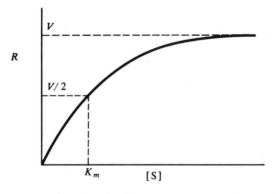

FIGURE 7-3   *Dependence of a steady state enzyme reaction rate R on substrate concentration [S].   The initial linear increase and eventual approach to a constant rate appear in all steady state treatments regardless of the number of intermediate species assumed.   In this drawing the parameters V and K_m refer to the constants that arise in the simplest form of the Michaelis–Menten mechanism,* $E + S \rightleftharpoons ES \rightarrow P$.

range 6 to 9, with the maximum value of V appearing at pH $= 7.8$. At pH $= 7$ and pH $= 8.5$ the value of V falls off to $\frac{1}{4}$ of the maximum value. Such results suggest that acid–base catalysis operates in enzyme reactions, and that the order of reaction with respect to H^+ and/or OH^- may range as high as 2. A third interesting general result of steady state enzyme kinetics experiments is that some molecules bind to the active sites of enzymes without undergoing any reactions at the site. Such species are called *inhibitors*. Inhibition of enzyme reactions proves to be a useful experimental artifice not only in steady state experiments, but also in the fast reaction and crystallographic experiments we discuss next. Further information of importance is obtained by analyzing the temperature dependence of steady state enzyme reaction rates in terms of activated complex theory. Enthalpies of activation range from about 1 kcal, typical of diffusion controlled reactions for example, to 20 kcal and more. Entropies of activation provide valuable hints about the nature of the interaction between substrates and active sites. The entropy of activation values for breakdown of enzyme–substrate complexes are large and *negative* in spite of the fact that we expect stretching of bonds to occur and hence a decrease of order. This suggests that interactions between an enzyme–substrate complex and the solvent are important in the energetics of enzyme catalysis.

By far the most important over-all contribution of steady state enzyme kinetics experiments, however, is the large body of general information about enzyme reaction mechanisms that emerges from them. Although the mechanisms inferred from steady state experiments are likely to be incomplete in almost all cases, as we shall see shortly in discussing fast reaction enzyme kinetics, they do provide an excellent framework into which the results of fast reaction and crystallographic experiments can be fitted. Further details of steady state experiments and a starting point for reaching the vast literature of the field of enzyme reaction mechanisms can be found in the Supplementary References listed at the end of the chapter.

Fast reaction enzyme kinetics. The study of enzyme kinetics at the steady state can give of course no information at all about the rates, and very little information about the mechanisms, of the elementary reactions responsible for establishment of the steady state. Steady state experiments can also give only indirect information about other elementary reactions which have rates that are substantially slower than the rate limiting reactions at the steady state. This limitation of steady state experiments was recognized early in the history of enzyme kinetics, and the desire of enzyme chemists to pursue the kinetics of fast reactions imbedded in enzyme reaction mechanisms was a key factor in the development of instrumentation for fast reaction studies. The technology of fast reaction experiments is discussed at length in Chapter 9. Here we are concerned more with the results of fast reaction experiments than with the methods, and the reader is referred to Chapter 9 and to the

Supplementary References at the end of Chapter 9 for further discussion of how the experiments are actually done. It will suffice here to state that two of the fast reaction methods have been especially fruitful in enzyme kinetics experiments. The first is the group of *flow methods* discussed in Section 9-1. The technology of flow systems for studying solution reactions was developed mostly by chemists studying enzyme reactions. The initial establishment of the steady state in enzyme reactions was studied for the first time in flow systems, and a wealth of important information on the transient behavior of enzyme reactions approaching the steady state was subsequently obtained. There is an inherent limitation of flow methods set by the maximum mixing rate that can be achieved, however, and this limits the reaction rates that can be studied to those which proceed on a time scale of milliseconds or longer. To investigate faster processes, the technique of *relaxation spectroscopy* (Section 9-3) is employed, principally in the temperature-jump version. In the usual apparatus employed for temperature-jump experiments the reaction times may be in the microsecond range, and thus three orders of magnitude faster than in flow systems.

The essential feature of fast reaction studies on enzyme reactions is that, in principle, the kinetic details of practically all of the steps of enzyme reactions can be obtained. In terms of the language of relaxation spectroscopy, each of the steps in an enzyme reaction is associated with a characteristic relaxation time, which can, in principle, be measured directly in an appropriate experiment. The complementary limitation of a relaxation spectroscopy experiment is that the identity of the elementary reaction responsible for each measured relaxation time must be inferred from other evidence, such as steady state experiments, pH effects, or general chemical knowledge.

Through temperature-jump and flow experiments it has been found that the number of intermediate stages in enzyme reactions can be very large. Two enzyme reactions that have been quite thoroughly studied with relaxation methods in particular are the splitting of ribonucleic acids (or model substrates similar to ribonucleic acids) by ribonuclease and the transfer of an amino group from aspartic acid to ketoglutaric acid by the enzyme aspartate aminotransferase. It was inferred from the combined results of studies of relaxation spectroscopic experiments and steady state experiments that at least *eight* states of the enzyme are involved in the mechanism of the first reaction and at least *fifteen* are involved in the mechanism of the second. It is clear that the mere pair of kinetic parameters provided by an analysis in terms of the Michaelis–Menten mechanism can represent a composite of the characteristics of a large number of elementary reactions indeed.

The finding that enzyme reaction mechanisms have a large number of discrete intermediate steps provides a great stimulus to the enzyme chemist to identify these steps with specific structural or chemical changes of the enzyme–substrate complex. Many of the suggested changes involve the conformation of the complex as the reaction proceeds. Other important contributions are certainly due to acid–base reactions that accompany the binding of the substrate molecules to the active site and the unbinding of the product molecules from the active site. Valuable hints concerning the acid–base reactions are provided by extensive fast reaction studies that have been made on nonenzymatic acid–base reactions. By using these results as a guide it is possible to suggest the identity of amino acid residues that are being affected by acid–base reactions as the enzyme reaction proceeds. In order to develop ideas about possible conformational changes that might accompany enzyme reactions, it is of course highly desirable to know what the enzyme and enzyme–substrate complex conformations actually are. We shall return to this subject in a moment.

Fast reaction experiments have shown that diffusion controlled reactions, or nearly diffusion controlled reactions, often occur in enzyme reactions mechanisms. One characteristic appearance of almost-diffusion-controlled rates is in the initial bimolecular step of enzyme–substrate complex formation. Typical values for the second order rate constant would be in the range 10^7–10^8 liters mole^{-1} sec^{-1}. The occurrence of some lower rate constants for complex formation—the complex between aspartate aminotransferase and α-methylaspartate, for example, is only 1.2×10^4 liters mole^{-1} sec^{-1}—is interpreted as an indication of important steric effects. Since complex formation is generally somewhat slower than diffusion controlled, it is usually distinguishable in enzyme reaction relaxation spectra from the faster elementary reactions of protonation or hydrogen bond dimerization, which usually have rate constants greater than 10^9 liters mole^{-1} sec^{-1}.

X-ray crystallography of enzyme structures. In discussing catalysis of gas reactions at solid surfaces we remarked that the nature of the active sites at the catalyst surface was rather difficult to control by procedures of catalyst purification, and equally difficult to characterize by experiments. In the case of enzyme reactions it is also an experimentally arduous procedure to determine the molecular structure of the active site, but once accomplished, the result provides spectacular insights into the mechanism of enzyme action. The molecular structures of enzymes can be determined by the method of x-ray crystallography, provided that a substantial amount of preliminary information in the form of the amino acid sequence and a substantial amount of difficult and tedious interpretive work on the analysis of the x-ray data are contributed by chemists interested in the structure of a particular enzyme. Even the smallest enzymes have molecular weight over 10,000, and with so many atoms in the enzyme molecules it is simply an enormous task to find out where all of them are located in space. The task is, of course, facilitated by the fact that the distances between the atoms of a single amino acid do not change appreciably when the amino acid is assembled into a polypeptide enzyme molecule. Two small enzymes for which accurate structures have been determined are ribonuclease, which has 124 amino acid residues, and lysozyme, which has 129 acid residues. The ribonuclease and lysozyme structures have also been determined with inhibitors bound to the active sites. Some larger enzymes for which less complete structural information has been obtained include papain, carboxypeptidase, chymotrypsin, and trypsin.

The folding of a polypeptide chain to form an enzyme molecule leads to certain characteristic features of the molecular shape. The best known of these is the fact that the chain tends to wind itself up into a helix. However, more important for the interpretation of enzyme action is the *tertiary structure*, which results when the helices of the chain are folded together to make the final enzyme molecule. When this occurs, and only when this occurs, the catalytic action is made possible. By careful inspection of the tertiary structure, and guided by knowledge of the molecular structure of the substrates whose reactions are catalyzed by a given enzyme, it is possible to locate the active site.

Let us consider now the question of enzyme specificity. What characteristic of the tertiary structure could be responsible for the observation that a given enzyme can only catalyze a single reaction? An obvious answer is that the geometry of the enzyme is such that only one kind of substrate molecule can be accepted and bound to the active site. This idea is actually an old one, and is commonly expressed in terms of an analogy to the fitting of a key into a lock. Confirmation of the truth of the idea, however, was first achieved through x-ray crystallographic determinations of the structure of lysozyme bound to inhibitors with structures similar to the actual substrates acted on by lysozyme. In the case of lysozyme, the active site has the appearance of a crevice running across the middle of the molecule. The inhibitor, and by inference the substrate as well, is bound along the length of the crevice. A similar situation obtains for ribonuclease.

There have not been enough complete enzyme structure determinations to permit a wide survey of the conformations of active sites. From the evidence available it would appear that there are at least three distinct types of conformation: active sites in the form of crevices, shallow depressions, and pits. Each type of site is characteristic of the reaction that the particular enzyme catalyzes. Lysozyme and ribonuclease, for example, serve the function of cleaving long polymer chains. It is plausible to assume that their active sites are in the form of crevices because during the evolution of enzymes it turned out that crevice-like active sites were the best chain cleavers. Carboxypeptidase, on the other hand, removes the end residue from a long polymer chain and has a pit into which the end of the substrate chain can be bound.

The specificity of a particular enzyme is then quite likely to be due to the details of its conformation at the active site. There is a wide variety of possibilities because there are about twenty different amino acids that commonly make up proteins, and several types of binding that each of them could contribute in forming an enzyme–substrate complex.

Having concluded that formation of an enzyme–substrate complex proceeds by a specialized conformation at the active site, we can now inquire into the nature of the catalysis itself. At one time it was thought that enzyme catalysis occurred simply because the active site was a favorable location for an ordinary chemical reaction to occur. In this view the enzyme would merely provide the location and the substrate molecules would do the rest. This idea turns out to be erroneous. In fact, the enzyme itself undergoes numerous alterations in conformation during complex formation and during the reaction at the active site. Such changes are seen best in the available x-ray structure of inhibited enzymes, but they also make themselves known by their appearance in relaxation spectra.

It should be clear at this point that a great wealth of detailed understanding of the mechanisms of enzyme reactions is made accessible by combining the results of steady state, fast reaction, and x-ray crystallographic studies. Since the field of enzyme kinetics research, after many decades of difficult progress, is just beginning to cross the threshold of complete mechanistic understanding of the chemistry of life processes, we can look forward to a profound increase in knowledge in this important area of science in coming years.

EXERCISES

7-17 A straightforward extension of the Michaelis–Menten mechanism is to the case in which the product molecules can react to reform reactants

$$E + S \underset{k_{-1}}{\overset{k_1}{\rightleftharpoons}} ES \underset{k_{-2}}{\overset{k_2}{\rightleftharpoons}} E + P$$

Show that the steady state rate law for this mechanism is

$$R = \frac{(V_f/K_f) [S] - (V_r/K_r) [P]}{1 + [S]/K_f + [P]/K_r}$$

$$V_f = k_2 [E]_0 \quad V_r = k_{-1} [E]_0$$

$$K_f = (k_{-1} + k_2)/k_1$$

$$K_r = (k_{-1} + k_2)/k_{-2}$$

How should the rate data be plotted to obtain the parameters V_f, V_r, K_f, and K_r? Show that the equilibrium constant of the reaction is given by $K_{eq} = V_f K_r / V_r K_f$.

7-18   The conventional Michaelis–Menten mechanism has the shortcoming that the intermediate species EP is not included.   An extension to include it would be

$$E + S \rightleftharpoons ES \rightleftharpoons EP \rightleftharpoons E + P$$

Use the steady state approximation on both ES and EP to show that a rate law of the Michaelis–Menten form is also obtained for this mechanism.   Relate the constants of the rate law for this mechanism with the constants of the rate law obtained in Exercise 7-17.

7-19   The initial rate of ATP dephosphorylation by the enzyme myosin can be estimated from the amount of phosphate produced in 100 sec.   Use the following data and the graphical methods of Exercise 7-6 to evaluate K_m and V for this reaction at 25°C. $[Myosin]_0 = 0.040$ g/liter.

[ATP]	$[Phosphate]_{100}$	[ATP]	$[Phosphate]_{100}$
7.1	2.4	70	6.2
11	3.5	77	6.7
23	5.3	100	7.1

All concentrations are in micromoles per liter.

7-20   At 5°C, the value of V for $[myosin]_0 = 0.090$ g/liter was found to be 10×10^{-7} moles liter^{-1} sec^{-1}.   Calculate the entropy and enthalpy of activation for the step ES $\rightarrow$ E + P of the ATP dephosphorylation reaction (cf. Exercise 6-1 and the data obtained in Exercise 7-19).

SUPPLEMENTARY REFERENCES

The principles of surface chemistry are presented in

A. W. Adamson, *Physical Chemistry of Surfaces* (Interscience, New York, 1967).

Fundamental aspects of heterogeneous catalysis are described in detail in

P. G. Ashmore, *Catalysis and Inhibition of Chemical Reactions* (Butterworths, London, 1963).

E. K. Rideal, *Concepts in Catalysis* (Academic Press, New York, 1968).

Electrode kinetics are discussed by

G. Charlot, J. Badoz-Lambling, and B. Tremillon, *Electrochemical Reactions* (Elsevier Publishing Co., Amsterdam, 1962).

B. B. Damaskin, *The Principles of Current Methods for the Study of Electrochemical Reactions* (McGraw-Hill, New York, 1967).   Presents introductory descriptions of experimental techniques for studying electrode reaction mechanisms.

P. Delahay, *Double Layer and Electrode Kinetics* (J. Wiley & Sons, Inc., New York, 1965).

K. J. Vetter, *Electrochemical Kinetics* (Academic Press, New York, 1967). Chapter 3 of this text presents the theory underlying methods of deducing the connection between overvoltage and elementary reaction rates; in Chapter 4 a large number of applications are discussed.

The fundamental relationship between current and overvoltage was first proposed by

J. Tafel, *Z. Physik. Chem.* **A50**, 641 (1905).

The principles and numerous examples of steady state enzyme kinetics are presented in

M. Dixon and E. C. Webb, *Enzymes* (Academic Press, New York, 1964).
K. J. Laidler, *The Chemical Kinetics of Enzyme Action* (Clarendon Press, Oxford, 1958).
J. Westley, *Enzymic Catalysts* (University of Chicago Press, Chicago, 1969).

The use of fast reaction methods for study of enzyme reactions is described in several articles to be found in the collection

S. Claesson, Ed., *Fast Reactions and Primary Processes in Chemical Kinetics* (Interscience, New York, 1967).

and also in a review article

G. G. Hammes, "Relaxation Spectrometry of Enzymatic Reactions," in *Accounts of Chemical Research* **1**, 321 (1968).

A description of the molecular structure of enzymes is given by

R. E. Dickerson and I. Geis, *Structure and Action of Proteins* (Harper and Row, New York, 1969), Chapter 4.

Chapter 8

CHAIN REACTIONS

MOST MECHANISMS have elementary reactions involving species that are not ordinary molecules at all, but highly reactive entities such as atoms, free radicals, or carbonium ions.   Often the free valences represented by such species persist through a long series of reactive encounters in which the identity of the species carrying the free valence continually changes and in which a large number of reactant molecules are converted into product molecules during the lifetime of a single free valence.   Such reactions are said to have *chain mechanisms*.   The two common types of chain mechanism are known as *linear* chain mechanisms and *branched* chain mechanisms.

Consider the schematic mechanism

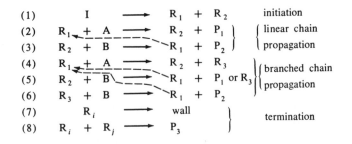

$$
\begin{array}{clll}
(1) & I \longrightarrow R_1 + R_2 & \text{initiation} \\
(2) & R_1 + A \longrightarrow R_2 + P_1 & \left.\right\} \left\{ \text{linear chain} \right. \\
(3) & R_2 + B \longrightarrow R_1 + P_2 & \left.\right\} \left. \text{propagation} \right. \\
(4) & R_1 + A \longrightarrow R_2 + R_3 & \left.\right\} \left\{ \text{branched chain} \right. \\
(5) & R_2 + B \longrightarrow R_1 + P_1 \text{ or } R_3 & \left.\right\} \left. \text{propagation} \right. \\
(6) & R_3 + B \longrightarrow R_1 + P_2 & \\
(7) & R_i \longrightarrow \text{wall} & \left.\right\} \text{termination} \\
(8) & R_i + R_j \longrightarrow P_3 & \\
\end{array}
$$

A and B are reactant molecules, the P_i are product molecules, and the R_i are *chain carriers* or *chain centers*.   The *initiation* step provides a starting source

212

of some R_i that can convert A and B into P_i by either of the two sets of *propagation* steps until they are removed from the scene of action by some kind of *termination* step. In terms of this schematic mechanism, various parameters characterizing the nature of any particular chain mechanism can be defined. The most important ones turn out to be the *chain length* of a linear chain reaction and the *branching coefficient* of a branched chain reaction. The former is defined as the average number of times reaction (2) occurs for each occurrence of reaction (1). The latter is defined as the average number of R_1 regenerated as the net effect of steps (5) through (8) for each occurrence of step (4), minus one.

We shall consider linear and branched chain mechanisms in terms of two well-characterized cases, the H_2–Br_2 reaction and the H_2–O_2 reaction.

8-1 LINEAR CHAINS: THE HYDROGEN–BROMINE REACTION

The gas-phase reaction between hydrogen and bromine to produce hydrogen bromide was shown by M. Bodenstein and S. C. Lind in 1906 to be a well-behaved homogeneous reaction over wide ranges of composition and temperature. They found, however, that it had a rate law which appeared peculiar:

$$R = k(T)\frac{([H_2]_0 - x)^{1.0}([Br_2]_0 - x)^{0.5}}{5.0 + x/([Br_2]_0 - x)^{0.5}}$$

where $x \equiv [HBr]$. Many years later a linear chain mechanism was found that predicts this rate law:

$$Br_2 \rightleftharpoons 2Br \tag{1}$$

$$Br + H_2 \rightleftharpoons HBr + H \tag{2}$$

$$H + Br_2 \rightarrow HBr + Br \tag{3}$$

Bodenstein and Lind had recognized the participation of bromine atoms from the square root dependence on bromine concentration, and had also proposed that HBr was formed by step (2). However, they thought the H atoms formed would recombine to form H_2. Overlooking the possibility that the H atoms produced in step (2) might react with Br_2 to form HBr and regenerate Br atoms prevented them from deriving their rate law from the mechanism; overlooking the possible reverse reaction of step (2) prevented them from understanding the retarding influence of HBr. Instead, they were diverted from the first discovery of a chain mechanism into a fruitless search for other reactions to explain the source of the HBr inhibition.

EXERCISE 8-1 Use the steady state assumption on [H] and [Br] to derive the Bodenstein–Lind rate law for the H_2–Br_2 reaction. (*Hint:* Add your two steady state equations together.) Identify the experimental $k(T)$ and 5.0 parameters with rate and equilibrium constants for the elementary steps (1)–(3).

The Columbus egg for deriving the experimental rate law was recognizing that a single free valence, associated alternately with H atoms and Br atoms, could lead to the production of a large number of HBr molecules at a rate that would change in time only through the effects of changes in the reactant and product concentrations. It was apparently not noticed in 1906 that a chain reaction could be a stable process in which the chain center concentration would be held at a constant value by the compensating rates of initiation and termination. Linear chain mechanisms, such as this one, which possess this characteristic are called *stationary* or *steady state* chains.

Some linear chains, it should be noted, have rates that increase in time, usually owing to the effect of a temperature rise caused by the heat being released in the reaction more rapidly than it can be dissipated to the environment. Three things can happen in such cases. The first is that the temperature of the reaction vessel eventually stabilizes at a point where the rate of heat loss to the environment is equal to the heat production rate. The second is that the reactants become depleted while the temperature is still rising, and the reaction turns itself off. The third possibility is a thermal explosion. As the temperature rises, the reaction rate increases, further accelerating the heat release, until the rate becomes catastrophic and the system explodes. The differential equations describing thermal explosions contain the effective activation energy, the exothermicity of the reaction, the heat conductivity, and the system size as principal parameters. By analyzing the differential equations describing the general system of an exothermic reaction in a vessel whose walls are kept at some constant temperature, the size of system at which eventual thermal explosion is inevitable can be calculated. The results of such critical size calculations for even very slowly self-heating systems such as moist flour are disconcertingly small. Less well defined but more familiar cases involve haystacks, piles of oily rags or the shipload of ammonium nitrate that thermally exploded at dockside in Texas City, Texas in 1947. Although there is no special reason for supposing that thermal explosions are restricted to linear chains, many of them turn out to be such.

Additional mention of the H_2–Br_2 reaction is made in Chapter 10 and in Exercises 2-2 and 8-7.

8-2 BRANCHED CHAINS: THE HYDROGEN–OXYGEN REACTION

This prototype branched chain reaction can be described for most conditions under which it has been studied by a mechanism that, although not simple by any definition, is readily understandable. However, instead of being characterized by the experiments of a single doctoral dissertation as was the

H_2–Br_2 reaction, the kinetics of the H_2–O_2 reaction have been the object of many experimental investigations.

It is convenient to describe the principal phenomena of the H_2–O_2 explosion in terms of the diagram shown in Figure 8-1. The graph shows the locus of a boundary in the p-T plane dividing p, T conditions under which a stoichiometric H_2–O_2 mixture will explode spontaneously from p, T conditions under which the same mixture will react very slowly or not at all. Consider the mixture at $p = 1$ torr, $T = 750°K$. Some atoms and radicals may be formed at the vessel wall or homogeneously, but they cannot lead to an explosion because at this low pressure their rate of diffusion to destruction at the vessel wall is greater than the rate of the chain propagation reactions. If the pressure is increased, the diffusion rate decreases while the chain propagation rate increases. At the *first limit* condition the latter dominates and the mixture explodes. Now consider a stoichiometric mixture at $p = 100$ torr, $T = 750°K$. No explosion occurs because radicals or atoms that may be present recombine in termolecular steps more rapidly than they react in bimolecular branching chain steps. If the pressure is lowered, the recombination rates are reduced more rapidly than the branching rate. At the *second limit* condition the latter predominates and the mixture explodes. Finally, if the $p = 100$ torr, $T = 750°K$ mixture is compressed, heat is generated in the slow reaction taking place under domination of the recombination steps more and more rapidly compared to its essentially constant rate of dissipation by conduction or convection. At the *third limit* condition, the self-heating leads to a thermal explosion.

Experiments carried out under conditions within the explosion region of the diagram show that the explosion is not instantaneous but follows a quiescent interval whose duration depends on the distance of the p, T point from the limits. This interval is called the *induction period*.

The following mechanism accounts for these facts:

initiation $H_2 + O_2 \rightleftharpoons HO_2 + H$ $\qquad\qquad\qquad$ (0)

$$H + O_2 \rightleftharpoons OH + O \qquad\qquad (1)$$

branching $\qquad O + H_2 \rightleftharpoons OH + H \qquad\qquad (2)$

$$OH + H_2 \rightleftharpoons H_2O + H \qquad\qquad (3)$$

$$H + O_2 + M \rightleftharpoons HO_2 + M$$

recombination $\left\{ H + OH + M \rightleftharpoons H_2O + M \right\}$ $\qquad$ (4)

$$H + H + M \rightleftharpoons H_2 + M$$

diffusion $H, OH, O, HO_2 \rightleftharpoons wall$ $\qquad\qquad\qquad$ (5)

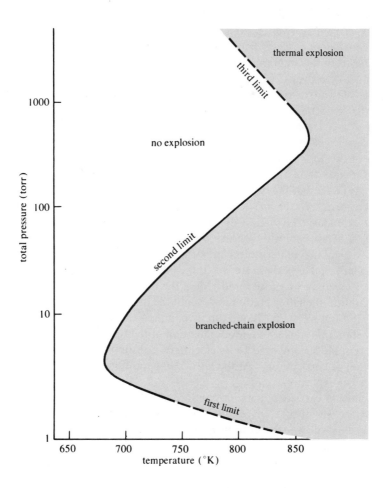

FIGURE 8-1 *Explosion limits for a stoichiometric mixture of hydrogen and oxygen. Mixtures whose pressure and temperature conditions lie well to the left of the line are stable. Mixtures whose pressure and temperature conditions are near the line but still to the left of it react slowly to form water and hydrogen peroxide. Mixtures whose pressure and temperature conditions are to the right of the line will explode. The locations of the first and third explosion limits are found to depend on the nature of the vessel in which the experiment is performed, whereas the location of the second limit does not. The second limit is affected by **addition** of inert gases, however. All of these facts have been rationalized in terms of a consistent mechanism. Quantitative measurements of the location of the explosion limits and of the reaction rates near the explosion limits have yielded rate constant values for several of the elementary reactions involved in the mechanism.*

In order for any reaction to take place at all, chain centers must be created, either homogeneously by (0) or heterogeneously by (-5). If (p, T) is below the first limit of explosion, they are lost to the wall more rapidly by (5) than they react homogeneously. If (p, T) is between the second and third limits, they recombine in reactions (4) more rapidly than they participate in the branched chain (1)–(3). If (p, T) is within the explosion "peninsula" between the first and second limits, the branched chain (1)–(3) takes over the process of chain center generation. To see how this behaves, form the sum $(1) + (2) + 2 \times (3)$:

$$H + 3\,H_2 + O_2 \rightarrow 2\,H_2O + 3\,H \qquad (1, 2, 3)$$

Each H atom reacting in (1) leads to generation of 3 H atoms in the subsequent steps (2) and (3). These 3 H atoms produce 9 H atoms in the next cycle; these 9 produce 27 in the next cycle, and so on. Eventually, regardless of how low the starting concentration of chain centers is or what their molecular identity may be, the branched chain will lead to gross consumption of H_2 and O_2 and an explosion will occur.

EXERCISE 8-2 Within the explosion peninsula of Figure 8-1 the rate controlling step of the over-all branched chain (1,2,3) is reaction (1), which has the rate constant expression $\log_{10} k_1 = 11.04 - 16060/4.57T$ in liters mole^{-1} sec^{-1} units. (a) During all but the very last part of the induction period, $[H_2]$ and $[O_2]$ remain constant. Show that under these conditions, if $[H] = [H]_0$ at time zero, $[H]$ at time t is given by $[H] = [H]_0 \exp(t/\tau)$, where the time constant τ is given by $\tau^{-1} = 2k_1[O_2]$. (b) Calculate the length of the induction period in a stoichiometric mixture for $p = 10$ torr, $T = 750°K$, assuming that there is one H atom in a 1.0-liter vessel at time $t = 0$, and defining the induction period to be over when $[H] = 0.001\,[H_2]_0$.

EXERCISE 8-3 Near the second limit of explosion in mixtures with $[O_2] \gg [H_2]$ the derivative $d\,[H]/dt$ is determined by the competition between $H + O_2 \rightarrow OH + O$ (1) and $H + O_2 + M \rightarrow HO_2 + M$ (4a). The second limit of explosion can be defined for this case as the value of p_{O_2} at which $\tau^{-1} = 0$. (a) Find a formula for τ as a function of k_1, k_{4a} and $[O_2]$. (b) Use the expression $\log_{10} k_{4a} = 9.0 + 1280/4.57T$ to calculate the value of p_{O_2} at the second limit of explosion in $[O_2] \gg [H_2]$ mixtures at $750°K$.

The equilibrium composition under the conditions shown in Figure 8-1 is virtually all on the water side of $2\,H_2 + O_2 \rightleftharpoons 2\,H_2O$. A moment's consideration will reveal that the branched chain steps (1)–(3) cannot produce this stoichiometry. The termolecular steps (4) can, however, and in fact they do.

Above the third limit of explosion, the mechanism is simply a coupling of the process of heat transfer out of the reaction vessel with the branched chain reactions (1)–(3), the termolecular reactions (4), and a few secondary atom and free radical reactions which we omitted from the mechanism presented above.   The result is a self-heating effect and a thermal explosion similar to those mentioned in the preceding section.

Branched chain mechanisms, although not as common as linear chains, do occur frequently in oxidation reactions.   It should be pointed out that branched chains do not necessarily lead to explosions, or in flowing systems to flames.   There are (p, T) regimes for all branched chain reactions in which the chain center concentrations have stationary values.   This can occur because there are reactions destroying chain centers that are first order in each of two chain centers or second order in one chain center, as the Br-atom recombination step in the linear chain of the H_2–Br_2 reaction.   In these cases the net rate of destruction will increase more rapidly with increasing chain center concentration than the net rate of branching, and a balance may be reached while the chain center concentration is still very low.   Another possibility is that some product of the branched chain may react efficiently with chain centers to remove them from the system, thus leading to *self-inhibition* of the reaction.

8-3   FREE RADICAL CHAINS

The mechanisms of the H_2–Br_2 and H_2–O_2 reactions consist of elementary steps that create, transform, or destroy chain centers in bimolecular or termolecular events.   Formally, we can regard the chain centers simply as atomic or radical species that undergo a linked series of reactions having the incidental effect of converting reactant molecules into product molecules from the moment a pair of centers is created in an initiation step to the moment a pair is destroyed in a termination step.   The essential difference between chain and nonchain, or *molecular*, mechanisms is the freedom given to the chain center by its *odd*, or *unpaired*, electron, which we can call its *free valence*.   Whenever an ordinary molecule reacts with a species containing a free valence, one of the products must also have a free valence; hence the chain center is conserved. Only when two chain centers react with one another, or when a chain center reacts heterogeneously at the surface of the reaction vessel, or when the free valence ends up on a species that for some reason is unreactive, can the chain center disappear.   Thus we recognize that if we are considering a chemically reactive system in which it is possible to form any species with a free valence, this free valence is likely to be conserved in subsequent elementary steps and generate a chain reaction.   Now we must raise the question of the importance of such free valence chains to chemical reactions generally.   In the H_2–Br_2 and H_2–O_2 reactions the rate laws themselves indicate chain mechanisms. For other free valence chain reactions, the experimental rate laws might just as well indicate molecular mechanisms, as we shall see in a specific example below.   Separate experiments in addition to the conversion rate measurements are needed to show the existence of free valences and free valence chains. To answer the question as to the importance of chain reactions first, however, the fact is that they seem to be just as prevalent as nonchain mechanisms.

The very existence of an unpaired electron on an atom or free radical automatically guarantees that its identity and concentration can be found, if only the concentration is large enough. In a magnetic field the two spin states of any unpaired electron have different energies, and radiation will be absorbed at the frequency satisfying the resonance condition $hv = \Delta \varepsilon$ (spin states). This *electron spin resonance* (ESR) method has enabled the direct observation of many atoms and free radicals.

Unfortunately for the study of reaction mechanisms, the unpaired electron concentrations needed for applying the ESR method are usually far greater than those which prevail during free valence chain reactions. Therefore, it is necessary to rely on indirect methods for detecting and identifying them. Although stable molecules with unpaired electrons have been known for many years, as for example nitric oxide and triphenyl methyl, which was discovered in 1900, it was a difficult task to show that reactive free valences participate in mechanisms generally. (For many particular cases it still is a difficult task.) Circumstantial evidence from photochemical experiments (Chapter 10) was reinforced in 1929 by the invention of the *mirror removal* experiment, shown in Figure 8-2. Metallic mirrors can be deposited on the walls of tubes containing vapors of the corresponding metal alkyls merely by heating the tube. Alternatively, the metal itself may be directly evaporated.

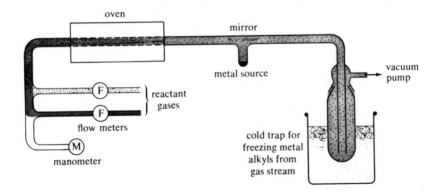

FIGURE 8-2 *The mirror removal experiment. The essential features of the apparatus shown schematically here are very similar to the flow system shown in Figure 9-1. The important difference is that the metallic mirror in the apparatus shown here can react with radicals present in very low concentration to give products that can be identified after the experiment is over by conventional analytical techniques. Mirror removal experiments are mostly of historical interest now, since fast reaction methods are available to study free radical reactions directly.*

If the tube is subsequently used as a flow reactor with a hot reaction section upstream of the mirror, free radicals flowing away from the reaction zone are found to remove the mirror by forming new metal alkyls.   In favorable cases, these alkyls can be frozen out of the stream of reaction products and analyzed to identify the radicals that removed the mirror.

After the presence of organic free radicals in a variety of thermal decomposition reactions had been shown by the mirror removal technique, there was no longer any reason to suppose that free radical mechanisms were uncommon.   A large number of techniques, some of which are mentioned in Section 8-5 and in Chapters 9 and 10, have since been devised for studying free radicals and free radical chains.

It remains to show here that chain reactions can have rate laws which are the same as rate laws for simple molecular mechanisms for the same reactions. A clear example for this is the thermal decomposition of ethane into ethylene and hydrogen

$$C_2H_6 \rightarrow C_2H_4 + H_2$$

which has the experimental rate law

$$R = k_{exp} [C_2H_6]^{1.0}$$

One might conclude from this rate law that the mechanism and the stoichiometric equation are identical, and that this is a unimolecular decomposition at its high pressure limit.   However, the mirror removal experiment shows that radicals are present.   Consider, then, the following chain mechanism:

$$C_2H_6 \rightarrow 2\,CH_3 \cdot \tag{1}$$

$$CH_3 \cdot + C_2H_6 \rightarrow CH_4 + C_2H_5 \cdot \tag{2}$$

$$C_2H_5 \rightarrow C_2H_4 + H \tag{3}$$

$$H + C_2H_6 \rightarrow C_2H_5 \cdot + H_2 \tag{4}$$

$$H + C_2H_5 \cdot \rightarrow C_2H_6 \tag{5}$$

The steady state assumption for H, $CH_3 \cdot$, and $C_2H_5 \cdot$, respectively, yields the equations

$$k_3\,[C_2H_5 \cdot] - k_4\,[H] - k_5\,[H][C_2H_5 \cdot] = 0$$

$$2\,k_1\,[C_2H_6] - k_2\,[CH_3 \cdot][C_2H_6] = 0$$

$$k_2\,[CH_3 \cdot][C_2H_6] - k_3\,[C_2H_5 \cdot] + k_4\,[H][C_2H_6] - k_5\,[H][C_2H_5 \cdot] = 0$$

Summing these three equations yields

$$[H] = \frac{k_1}{k_5} \frac{[C_2H_6]}{[C_2H_5\cdot]}$$

Substituting this result into the steady state equation for H yields the quadratic equation in $[C_2H_5\cdot]$

$$k_3 k_5 [C_2H_5\cdot]^2 - k_1 k_5 [C_2H_6][C_2H_5\cdot] - k_1 k_4 [C_2H_6]^2 = 0$$

which has the solution

$$[C_2H_5\cdot] = \frac{k_1 k_5 [C_2H_6] \pm \{(k_1 k_5 [C_2H_6])^2 + 4 k_1 k_3 k_4 k_5 [C_2H_6]^2\}^{\frac{1}{2}}}{2 k_3 k_5}$$

$$= \left\{ \frac{k_1}{2k_3} \pm \left[\left(\frac{k_1}{2k_3}\right)^2 + \frac{k_1 k_4}{k_3 k_5} \right]^{\frac{1}{2}} \right\} [C_2H_6]$$

This equation can be simplified if we assume $k_1 \ll k_3$ (cf. Exercise 8-5). Then for any reasonable k_4/k_5 ratio we obtain

$$[C_2H_5\cdot] = \left[\frac{k_1 k_4}{k_3 k_5} \right]^{\frac{1}{2}} [C_2H_6]$$

and the rate law

$$R = \left[\frac{k_1 k_3 k_4}{k_5} \right]^{\frac{1}{2}} [C_2H_6]$$

EXERCISE 8-4 Work out the derivation of the preceding paragraph in detail.

EXERCISE 8-5 (a) Estimate the values of the rate constants for this mechanism at 900°K using typical (Chapters 4 and 5) values for the preexponential factors and estimating activation energies from a table of bond energies or bond dissociation energies. (b) Verify the validity of the simplifications used to solve the above quadratic equation on the basis of your estimated rate constants. (c) Calculate the over-all activation energy, that is, the activation energy for k_{exp} from the rate law and your estimated activation energies for k_1, k_3, k_4, and k_5.

EXERCISE 8-6 Calculate the chain length at (1) 800°K and (2) 1,200°K for your estimated rate constants. *Hint:* Compare the rate of C_2H_6 decomposition by reaction (3) with the rate of decomposition by reaction (1).

EXERCISE 8-7 (a) Estimate the activation energies for the rate constants appearing in your answer to Exercise 8-1 by using a table of enthalpies of formation, and combine this information with your identification of the experimental rate constant with elementary reaction rate constants to find the activation energy for the initial stages of reaction ($[HBr] \ll [Br_2]$). (b) Estimate the chain length at 1,000°K.

Space limitation has prevented us from discussing the general topic of polymerization kinetics, which is of considerable practical importance because the properties of polymers are strongly dependent upon the kinetics of poly-merization. At this point, however, it is useful for us to consider the particular subject of *free radical polymerization*. Actually, only a fraction of the important industrial and biochemical polymerization reactions have such mechanisms; many of them have molecular mechanisms, such as the *addition* polymerizations that form polypeptides or polyesters, and many others have *ionic* chain mechanisms in which the freedom of the valence is generated by a Lewis-acid or Lewis-base catalyst. Of the polymerizations that do have free radical mechanisms, the polymerizations of olefins, such as ethylene, styrene, or vinyl chloride, are the most important.

The general mechanism for free radical polymerization can be written out as follows. Let M represent an individual molecule, or *monomer*, capable of undergoing polymerization and let R_j represent a polymeric radical con-taining j monomer units. The mechanism is

$$\text{initiation} \rightarrow R_1$$

$$R_1 + M \rightarrow R_2$$
$$\vdots$$
$$R_k + M \rightarrow R_{k+1}$$
$$\vdots$$
$$R_m + R_n \rightarrow R_m R_n \quad \text{(termination)}$$

Initiation can be thermal, photochemical, or catalytic, with the initiation rate having a corresponding dependence on temperature, light intensity, or catalyst concentration. Each chain propagation step then increases the length of the polymeric radical by one monomer unit. Termination occurs when two polymeric free radicals combine their free valences to form a saturated polymer molecule.

EXERCISE 8-8 (a) Write rate laws for free radical polymerization using each initiation mechanism mentioned, and making an assumption about the rate con-trolling step in each case. (b) Derive a formula that can be used to calculate the initiation rate from $d[M]/dt$ and the average molecular weight of the polymer.

8-4 EXPERIMENTAL TESTS FOR CHAIN REACTIONS

It is critically important in doing experiments on reaction mechanisms to determine whether chain reactions are occurring. Some of the principal ways to do this are listed here.

Detection of intermediates. In favorable cases the mirror removal or the ESR method can allow detection of free radicals. Another possibility is optical spectroscopy combined with the flash photolysis or shock tube techniques (Chapter 9). Free radicals have also been identified in gas reactions by removing gas from the reaction vessel into a sensitive mass spectrometer.

Sensitizer yield. Some unstable molecules can be decomposed thermally or photochemically to produce known quantities of known radicals. An example would be azomethane, CH_3—$N{=}N$—CH_3, which decomposes to form methyl radicals. Studying the effects of adding a known radical source can allow one to make a decision on the nature of the unperturbed reaction.

Isotopic scrambling. Synthetic methods exist for making molecules in which some or all atoms are *labeled* with special isotopes, for example, deuterium or tritium, carbon-14, nitrogen-15, or oxygen-18. If such labeled molecules are used in place of normal reactants, the distribution of the isotope in the product molecules can allow a distinction to be made between chain and nonchain mechanisms. This method must be combined with kinetic measurements to give definitive results.

Effect of inhibitors. A number of substances are known to react rapidly with certain atoms or free radicals to form unreactive products and hence terminate chains. Iodine, for example, removes H atoms by forming HI and I atoms, both of which are unreactive compared with H atoms under most conditions. If trace additions of such *inhibitors*, or *scavengers*, cause strong suppression of a reaction, a free radical chain is indicated.

Induction periods. It is possible that nonchain mechanisms can show induction periods caused by the thermal self-acceleration effect discussed in Section 8-2, or other effects. The well-known iodine clock reaction is a case in point. For the most part, however, induction periods indicate chain reactions.

Wall effects. Atoms or free radicals can be destroyed when they diffuse to the walls of a reaction vessel; the efficiency of the destruction depends on the type of wall surface. If a reaction proceeds at different rates in vessels of different size or different composition, a free radical chain is indicated. This effect can first appear when the measured conversion rates appear to fluctuate greatly under apparently identical experimental conditions, with the fluctuations being attributed to changes at the vessel surface.

Product distribution. A complete analysis of the reaction products can provide strong inference that a chain mechanism prevails; it may simply be impossible to write a nonchain mechanism converting the known reactants to the known products. The converse, it must be stressed, does not hold at all; simple rate laws and/or simple product distributions do not imply molecular mechanisms, as we saw in the ethane example of Section 8-3.

Acceleration of a known chain reaction. In some cases, free radicals can be detected by their known participation in certain chain reactions. The best known example is the *ortho–para* hydrogen conversion, which proceeds by the mechanism $H + o\text{-}H_2 = p\text{-}H_2 + H$ at an accurately known rate. If a small addition of p-H_2 isomerizes in a reaction mixture, free radicals are surely present.

Activation energies.   We found in Chapter 4 that elementary reaction activation energies are not less than reaction heats for endothermic elementary reactions.   If observed activation energies are lower than those expected for the molecular mechanisms compatible with the rate law and the product distribution for a reaction, a free radical chain may be involved.   An example of this situation was presented in Exercise 8-5.

Quantum yields.   In Chapter 10 we discuss the subject of photochemistry, comprising the study of reactions that occur upon absorption of radiation.   If the ratio of molecules produced to quanta absorbed, called the *quantum yield*, is a large number, chain reactions are involved.

We conclude our listing of diagnostics for chain reactions by remarking again that really conclusive experiments are rare in the study of reaction mechanisms and that understanding them always comes as a result of reconciling the results of many experiments.

SUPPLEMENTARY REFERENCES

Chain reactions have been discussed in extensive detail by many authors. Some well-known references are listed here.   The books by Dainton and Trotman-Dickenson are particularly good introductions to the subject.

J. Bevington, *Radical Polymerization* (Academic Press, Inc., New York, 1961).

F. S. Dainton, *Chain Reactions* (J. Wiley & Sons, Inc., New York, 1956).

G. J. Minkoff and C. F. H. Tipper, *Chemistry of Combustion Reactions* (Butterworths, London, 1962).

N. N. Semenov, *Chemical Kinetics and Chain Reactions* (Oxford University Press, Oxford, 1935).

N. N. Semenov, *Some Problems in Chemical Kinetics and Reactivity* (Princeton University Press, Princeton, N.J., 1958, 1959), Vols. I and II.

A. F. Trotman-Dickenson, *Free Radicals: An Introduction* (Methuen, London, 1959).

Chapter 9

FAST REACTIONS

MOST REACTION mechanisms presume the participation of some elementary reactions that are too fast for study under experimental conditions generally used in chemistry and that involve intermediate species whose existence is only inferred indirectly or whose reactivities are so great that their concentrations are too low for detection. The traditional attitude of the chemical kineticist to this situation was one of resignation: Since mechanisms are only hypotheses anyway, there is no fundamental objection to their containing convenient but unconfirmable ephemeral events. On the other hand, we can envision such a thing as an ideal mechanism—a mechanism that provides exact qualitative and quantitative rationalization of all experimental facts while including only known chemical species and elementary reactions with known rates. The degree to which any mechanism approaches this ideal is one measure of the confidence that can be placed in its correctness. In order that the science of reaction mechanisms be advanced, it is desirable that the rates of as many elementary reactions as possible be studied individually, and the characteristic reactivities of common intermediates be determined, so that actual mechanisms may come closer to the ideal. To achieve this, the rates of the elementary reactions that are "instantaneous" under ordinary conditions must be known.

Before discussing the problems encountered in studying fast elementary reactions we must decide what we really mean by "fast." Large values of the rate constant alone will not provide a criterion, for rate constants can be changed by altering the temperature, and conversion rates can be changed by altering the reactant concentrations. Thus, unimolecular decompositions that occur on the time scale of molecular vibrations or bimolecular reactions that occur at every encounter are obviously fast as far as the molecular changes are concerned, but the corresponding conversion rates can be arbitrarily small if the rate of activation of the decomposing molecules or the encounter frequency are reduced. We shall find it useful to adopt the pragmatic definition that a fast reaction is one which has a half-time that is small compared with the response time of humans, for example on the order of seconds, when the reactants are present in concentrations usually employed for chemical reactions (e.g., 0.1 molar) and the temperature is room temperature or above. In adopting this definition we include reactions with complicated mechanisms as well as isolated elementary reactions; however, the similarity in experimental techniques for the two cases makes it useful to carry through with this definition. Study of these fast mechanisms turns out to be an especially fruitful extension of conventional methods, for many of the most important mechanisms in chemistry are fast by this definition.

The essential difficulty of studying the kinetics of fast reactions, which for many decades limited the chemical kineticist to studying reactions with half-times at least in the seconds range, lies in two complementary limitations of analytical chemistry. Analytical methods fast enough to follow rapid reactions are ordinarily insensitive; methods sensitive enough to detect low concentrations are ordinarily slow. In order to follow fast reactions we must circumvent or overcome this characteristic of chemical analysis.

For reactions with appreciable activation energy, the analysis problem can be circumvented in some cases simply by reducing the temperature. For example, a reaction with an activation energy of 20 kcal will be 10^7 times slower at $157°K$, the freezing point of ether, than at room temperature. By measuring the Arrhenius parameters at low temperatures, the rates of reactions that have half-times of small fractions of a second at higher temperatures can be calculated. A second device that also amounts to circumvention of the analysis problem is to take advantage of certain extraordinarily sensitive analytical procedures and carry out the rate measurements in extremely dilute solutions. For a second order reaction, the conversion rate with 10^{-6} molar concentrations of both reactants is clearly 10^{10} times slower than if their concentrations are 10^{-1} molar. Unfortunately, both of these routes of circumvention, and others that have been devised as well, appear to be restricted to just a few favorable cases; for when such large changes in the reaction conditions are made, the mechanism usually changes also.

One means of circumventing the analysis problem that is generally useful for the study of fast reactions with equilibrium constants much greater than unity is to measure the much slower reverse rate and calculate the forward rate constant from $k_f = K_{eq} \times k_r$. Examples of this procedure have been given in connection with diffusion controlled reactions (Section 6-5).

Another means is to establish a stationary state in the system and observe the interrelation of fast chemical reactions with fast physical processes such as diffusion or light emission. Diffusion controlled electrode reactions (Section 7-2) and fluorescence quenching (Section 10-2) are successful applications of this idea. In each case there is a fast internal time scale, provided by the diffusion constant in the first case and the fluorescence lifetime in the second, which permits very fast chemistry to be followed at the stationary state condition.

The most fruitful way of overcoming the analysis problem is to overcome its time response aspect, and it is partially on this account that the title "fast reaction" is given to this field of study. In fact, efforts to speed up analysis times have been so successful that new methods of initiating chemical reactions had to be invented to allow the study of reactions that are faster than reactant mixing. Reaction half-times as short as 10^{-9} sec have been measured.

The idea of speeding up analytical methods is obvious, but realizing the required speedup in practice presents a severe challenge. It has been accomplished through the great advances in technology, primarily in electronics, that have occurred since World War II. Progress in the study of fast reactions has essentially paralleled progress in measurement technology. The resulting increased understanding of reaction mechanisms and large number of elementary reaction rate constants have revolutionized chemical kinetics as thoroughly as the new technology has revolutionized electrical engineering. Continuing progress in measurement technology will, however, increase the understanding of fast reactions more in breadth than in depth; for we shall find that kineticists are already able to follow transformations which are so fast that the distinction between chemical processes (i.e., making and breaking chemical bonds) and physical changes (e.g., molecular rotations or exchange of energy in collisions) becomes so blurred that it is meaningless.

In this chapter we discuss the experimental principles underlying six methods of studying fast reactions. Two other methods are discussed elsewhere in this book—fluorescence quenching in Section 10-2 and diffusion controlled electrode reactions in Section 7-2.

9-1 FLOW SYSTEMS

The most direct way to deal with the difficulty of making chemical analyses quickly is by running reactions in flowing systems. This was indeed the first fast reaction method invented, and it is still a very important one, capable of extending one's ability to observe reaction times by four orders of magnitude, from a few seconds to fractions of a millisecond. The basic idea is to flow the reactants together in a mixing chamber, from which they leave together and travel at high speed down a tube. From the measured flow rate, the elapsed time required for the reacting mixture to travel from the mixing chamber to any given point in the tube can be calculated quite simply, with the chain rule transforming the conversion rate from a time derivative to a space derivative, $d[A]/dx = (d[A]/dt)(dx/dt)^{-1}$. Chemical analyses can be done most directly by optical absorption spectroscopy, although other methods have also been used.

The time for analysis is limited only by the supply of reactants available to feed into the mixing chamber. Both gas and solution reactions have been studied with essentially the same techniques and the same degree of success. The main components of one flow apparatus suitable for studying fast gas reactions are shown in Figure 9-1.

The limitations that define the range of applicability of flow techniques to chemical kinetics are hydrodynamic or gas dynamic ones. When the flow speeds in the mixing chamber and in the flow tube become high, a point is eventually reached at which viscous losses and mixing imperfections become intolerable. These effects limit the fastest reaction half-times measureable in flow systems to about 10^{-3} sec. In studying gas reactions in flow systems it is necessary to keep the pressure high enough so the concentration changes are not grossly affected by diffusion as well as by reaction.

To illustrate the use of flow systems in chemical kinetics, we consider the theoretically important reactions of hydrogen atoms and hydrogen molecules, including the isotopic variants. Hydrogen or deuterium atoms can be generated in one of the gas flow lines by passing the gas through an electrical discharge. In most experiments the hydrogen or deuterium gas is present in a large excess of argon carrier gas. Among the reactions that have been studied are

$$D + H_2 \rightarrow HD + H$$

$$H + D_2 \rightarrow HD + D$$

$$H + H + Ar \rightarrow H_2 + Ar$$

The corresponding rate constants are listed in Table 4-1 and Figure 5-3. In various laboratories, H-atom concentrations were followed by such diverse methods as calorimetry, ESR spectroscopy, absorption spectroscopy of H atoms at 1216 Å, and mass spectroscopy. The flow tubes have been in thermostats covering temperature ranges as wide as 250 to 1,000°K. The flow tube pressure is typically a few torr. By measuring the flow velocity and the decay of atom concentration with distance, the rate constants for the reactions listed could be derived. The Arrhenius parameters are in approximate, but not exact, agreement with the results of ACT calculations (Section 4-7) and trajectory studies (Section 4-4). References to the original literature describing these experiments are given at the end of the chapter.

A variation of the flow method that can be applied to the study of slower reactions is the *stirred reactor* or *capacity flow* method. Here the reactants are added at a constant rate to a large stirred reaction vessel while products are removed at the same volume flow rate. Analyzing the product stream gives the composition within the reaction vessel itself, and a simple calculation (Exercise 9-1) connects the rate constants to the composition and the flow rates.

Two variations of the flow method that are useful for studying fast solution reactions are the *stopped flow* and *accelerated flow* techniques. In these variations the flow is suddenly brought to a halt or suddenly accelerated, after which the reaction progress at a fixed point in the tube is followed by high speed analytical methods such as kinetic spectroscopy (Section 9-4). These variations can incorporate the advantage of using small quantities of reagents, which may be critical in biochemical experiments. The stopped flow technique has the additional advantage that slow reactions can be studied with the identical apparatus as fast reactions. The kinetics of enzyme reactions have been investigated with great success using these techniques (Section 7-3).

Two special cases of flow experiments in gas kinetics are *premixed flames* and *diffusion flames*. The premixed flame experiment is a refined version of the Bunsen burner. It is usually run at reduced pressure since the reaction rates are so high that quite low concentrations are required to make the spatial resolution of the reaction zone compatible with available analytical

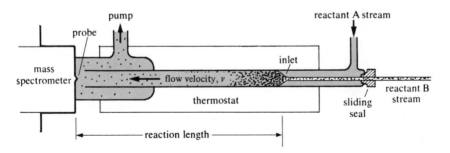

FIGURE 9-1 *Schematic diagram of a flow apparatus suitable for study of fast gas reactions. The mass spectrometer probe provides a quantitative and qualitative analysis of the reaction mixture at an equivalent reaction time $t = x/v$, where v is the flow velocity in the reaction tube and x is the distance between the mixing inlet and the probe. In experiments of this kind it is essential to consider the possible effects of diffusion, wall reactions, finite mixing rates, and reactions within the probe.*

methods. The geometry is made one-dimensional by flowing the premixed reactant gas either through a large, flat porous plate to give a virtually flat flame, or through a small orifice in a needle point to give a virtually spherical flame. In the diffusion flame, as the name implies, two reactant gases diffuse into one another at low (0.1–10 torr) pressure. The profiles of either type of flame are determined by temperature probes or by absorption and emission spectroscopy. In the case of the diffusion flame, one has the capability of making absolute measurements of reaction rates that occur at nearly every encounter, and it is possible to study isolated elementary reactions. In studying complex combustion mechanisms by means of premixed flames, the extraction of elementary reaction rate constants is complicated not only by the usual mechanistic difficulties but also by the difficulties that the temperature and the density change within the flame and that the diffusion rates of the species in the flame are often critically important and uncertain.

9-2 SPECTRAL LINE SHAPES

Atomic and molecular spectroscopy. In the potential energy curves of simple molecules (Figure 10-2) one normally represents the stationary vibrational– rotational states as having discrete energy values, with the transitions between levels involving photons of a single wavelength given by $\lambda = hc/\Delta E$, or a single frequency given by $v = h/\Delta E$. In reality, single frequencies are not observed in spectroscopic experiments because the emitting or absorbing molecules move with respect to the observer, because the energies of the emitting or absorbing molecules are perturbed by other molecules, even at low pressures in the gas phase, and because of other effects. Long ago physicists realized that graphs of absorptivity (cf. Section 10-1) versus frequency, which came to be called *line shapes*, were capable of yielding valuable information about the physical interactions of atoms or molecules with one another. Chemists are now using line shapes to derive information about reactive interactions of molecules as well.

The simplest chemically relevant information provided by line shapes concerns the exchange of energy in encounters. Energy transfer between translational motion and vibrational or rotational internal excitation was shown in Chapters 4 and 5 to play a fundamental role in the theory of elementary reactions; in Chapter 10 the exchange of translational and electronic energy is shown to play a role in photochemistry. The theory underlying the use of line shapes for determining the cross sections for various types of energy transfer is usually known as the theory of *line broadening* since the primary effect of molecular interactions is to widen the absorptivity versus frequency graph. Although by no means difficult, this theory is well beyond the scope of our book. We can only note in passing that cross sections for various types of energy transfer processes can be and are obtained by line shape experiments with all types of spectroscopy from uv ($v \cong 10^{15}$ sec^{-1}) to microwave ($v \cong 10^{10}$ sec^{-1}).

The line shape of a single line of an atomic spectrum is often described by a function of the form

$$a(v) = \frac{a(v_0)}{1 + [2(v - v_0)/\Delta v_L]^2}$$

where $a(v)$ is the absorptivity (cf. Section 10-1) at frequency v, and v_0 is the frequency of maximum absorption. This function came to be called a *Lorentzian* in honor of H. A. Lorentz, who first derived a theory of pressure-broadened line shapes of atomic spectra. The quantity Δv_L, which is the width of the graph where the absorptivity has $\frac{1}{2}$ its maximum value $a(v_0)$, is called the *Lorentz half-width*. Lorentz showed in 1906 that this quantity is related to the average time τ_L between encounters responsible for broadening in the gas phase according to

$$\Delta v_L = (\pi \tau_L)^{-1}$$

(The invention of quantum mechanics gave rise to a simple and instructive way to derive a relationship of this kind. One form of the uncertainty principle is

$$(\Delta E)(\Delta t) \geq h/4\pi$$

setting $\Delta E = \Delta(hv) = h\, \Delta v$ and using the equality yields

$$\Delta v = (4\pi \, \Delta t)^{-1}$$

which, except for the factor 4, is identical in form to the Lorentz line width formula relating Δv_L to τ_L. The broadening Δv is sometimes called "uncertainty broadening" and attributed to an increase in "energy uncertainty" that accompanies shortening the "lifetime" of absorbing (or emitting) atoms or molecules to the value Δt.)

The role of line shape experiments in the study of energy transfer is described in the photochemistry supplementary references at the end of Chapter 10 and in the energy transfer supplementary references at the end of Chapter 4. The variety of subjects treated in these references shows that energy transfer impinges on chemistry in a variety of ways. It might also give the incorrect impression that the theory underlying the various line shape experiments differs from one type of experiment to another. This is not the case at all. In each experiment one finds the average length of time that an atom or molecule spends in the quantum states observed in the transition.

If the two quantum states have energies E' and E'', then the energy of the transition is $\Delta E = E' - E''$, which corresponds to a frequency $v = h^{-1}(E' - E'')$. If the average lifetimes in the two states are denoted by τ' and τ'', then the Lorentz line width of the emission or absorption line is given by the formula

$$\Delta v_L = (\pi \tau_L)^{-1} = \pi^{-1}(\tau'^{-1} + \tau''^{-1})$$

In some cases, including the atomic absorption case considered by Lorentz, one of the lifetimes is much longer than the other, and the line width gives the lifetime of the short-lived state only.   This is a favorable situation for studies of energy transfer.   One measures the increase in line width when a gas is added to shorten the lifetime of the short-lived state and calculates a cross section or a bimolecular rate constant from the Lorentz line width formula (cf. Exercise 9-12).   Unimolecular processes can also be responsible for the lifetime of a quantum state; in this situation the average lifetime can be related to a specific unimolecular decay constant $k_a(\varepsilon)$.

An example of a line width experiment on energy transfer concerns rotational energy transfer in collisions of carbon oxysulfide (OCS) molecules. The microwave absorption line corresponding to the transition from rotational state $J = 1$ to rotational state $J = 2$ was found to have a Lorentz line width of $12.4 \times 10^6 \text{ sec}^{-1}$ at 1 torr pressure.   Conversion of this figure to a cross section, which is left as an exercise for the reader, shows that the cross section for transfer of rotational energy is about twice the hard sphere cross section derived from viscosity measurements (Exercise 9-15).

Magnetic resonance.   Information about reaction rates is provided quite directly by the special properties of magnetic resonance spectroscopy, both ESR and NMR.   In ESR and NMR experiments the magnetic energy levels of the unpaired electrons or the nuclei whose resonances are being observed function as sensitive indicators of the chemical environment of those particular electrons or nuclei.   Identification of these environments through the values of the chemical shifts and the hyperfine splittings is the primary application of magnetic resonance to chemistry.   The same splittings and shifts, however, also serve as excellent internal probes of the reaction rates affecting the chemical environments that are responsible for them.   The reaction rates that can be obtained from magnetic resonance spectroscopy range from 1 to 10^{-9} sec half time and therefore fall into our fast category. Although the reactions that can be studied and the range of rate constants that can be measured are different in ESR and NMR, the general features of the theory relating the line shapes to reaction rates are the same; thus we can discuss both types of magnetic resonance together.

Line shapes in ESR and NMR spectra of liquid solutions can often be described by Lorentzian functions.   In place of the absorptivity $a(v)$, which is appropriate for describing the decrease in intensity of a light beam passing through matter, the Lorentzian line shape in magnetic resonance spectroscopy describes *normalized lines shape functions* $g(v)$ or $g(H)$, where H is the applied magnetic field. The functions $g(v)$ and $g(H)$ are proportional to the amount of power absorbed at frequency v or magnetic field H due to a magnetic resonance absorption; the constant of proportionality is chosen to normalize the graphs

$$\int_{\substack{\text{absorption}\\\text{line}}} g(v)\,dv = \int_{\substack{\text{absorption}\\\text{line}}} g(H)\,dH = 1$$

The Lorentzian function for $g(v)$ is

$$g(v) = \frac{2T_2}{1 + 4\pi^2 T_2^2 (v_0 - v)^2}$$

The Lorentzian line width Δv_L is expressed in $g(v)$ through $\Delta v_L = 1/\pi T_2$, the quantity T_2 being a fundamental parameter in magnetic resonance theory known as the *transverse relaxation time*. The relationship between v and H, and hence between Δv_L and ΔH_L, depends on the magnetic species being studied. For electrons, $v\ (\text{sec}^{-1}) = 2.8 \times 10^6\ H$ (gauss); for protons $v = 4.26 \times 10^3 H$, for ^{19}F, $v = 4.01 \times 10^3 H$; for ^{17}O, $v = 5.77 \times 10^2 H$. Derivation of normalized line shape functions $g(H)$ is left as an exercise for the reader (Exercise 9-14).

The atomic spectroscopy relationship $\Delta v_L = (\pi \tau_L)^{-1}$ has several analogs in ESR and NMR spectroscopy. For a single hyperfine component, the quantity τ_L is called a *mean lifetime* and refers to the average time that an electron or a nucleus spends in the magnetic environment corresponding to that hyperfine component. If the width of a resonance line is entirely determined by the fact that a nucleus or an electron changes its environment by chemical reaction, then $\Delta v_L = (\pi \tau_L)^{-1}$ and finding Δv_L is completely equivalent to finding the reaction rate. Starting from this, one can immediately begin to carry out the usual chemical kinetics procedures of varying concentrations to determine reaction order, varying the temperature to determine activation energy, varying the solvent to study solvent effects, and so on. The usual case in magnetic resonance, however, is that Δv_L is affected by a number of other factors (magnetic field inhomogeneity and solute–solvent interactions are the main ones) besides chemical reaction rates. For this reason, serious systematic errors can arise in magnetic resonance kinetics experiments if all sources of line broadening are not properly taken into account.

The type of line width experiment just described is known as a study near the *slow exchange limit*. Consider now what may happen in a proton magnetic resonance study if the rate of exchange of a proton between two molecules

$$B + HA \rightleftarrows HB + A$$

is steadily increased by increasing the reactant concentrations or increasing the temperature.   As the rate of exchange increases, the mean lifetime τ_L decreases and the line width $\Delta v_L = (\pi \tau_L)^{-1}$ increases.   As the resonance lines broaden, their widths can become comparable to the chemical shift that separates the two resonances, as shown in Figure 9-2.   They begin to overlap and finally coalesce into a single line centered at the average resonance position.   Further increase of the exchange rate then leads to a *narrowing* of the merged proton resonance line.   When this occurs the proton is being exchanged so rapidly between the two molecules that the observed chemical shift is associated with the average magnetic environment in HA and HB rather than the magnetic environment in either of the species alone.   The line narrowing effect is known as approaching the *fast exchange limit.*   For a few favorable cases it has been possible to observe the effect of exchange upon the line width all the way from the slow exchange limit to the fast exchange limit.   In most rate studies by magnetic resonance, however, only part of the line narrowing or line broadening can be followed experimentally.

Since there is no analog in atomic spectroscopy to line coalescence or fast exchange, the theory of these effects had to be developed from the beginning for magnetic resonance spectroscopy.   When this was done it was found that the simple Lorentz theory is not adequate except for initial orientation as to the magnitude of the rates near the slow exchange limit.   Algebraically complicated, although conceptually still simple, theories are needed to give the correct connections between observed resonance line shapes and exchange rates.

In magnetic resonance theory the mean lifetime concept from the Lorentz theory is retained.   Our particular interest is in a quantity τ_A, the mean lifetime in magnetic environment A for exchange of a nucleus (in NMR) or an electron (in ESR).   The exchange of environment may result from a bimolecular reaction, such as an electron transfer reaction or the proton transfer reaction discussed above, or it may result from a unimolecular reaction such as a *cis–trans* isomerization.   For a bimolecular exchange reaction A + B $\rightleftarrows$ B + A the mean lifetime of A is related to the second order rate constant k^2 by

$$\tau_A = \frac{[A]}{k^2 [A][B]} = (k^2 [B])^{-1}$$

and the mean lifetime of B is similarly $\tau_B = (k^2 [A])^{-1}$.   For a unimolecular

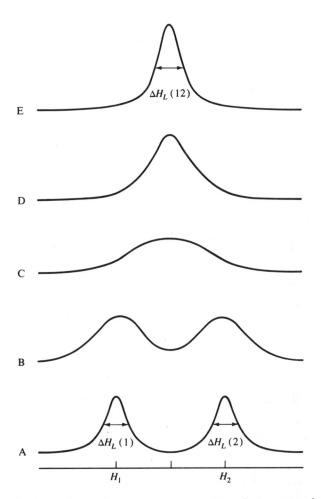

FIGURE 9-2   *Dependence of magnetic resonance line shapes on exchange rates. The case illustrated here is a proton exchange* $HA \leftrightarrows HB$ *for which* [HA] = [HB] *and* $T_2^0(A) = T_2^0(B)$. *In the slow exchange limit (A) the* HA *and* HB *proton resonances occur at different magnetic fields,* H_1 *and* H_2. *As the exchange rate is increased (B), the peak separation decreases and the lines begin to merge. Further increase in the exchange rate leads to coalescence (C) and then narrowing of the coalesced line (D and E). At the fast exchange limit (E) the chemical shift is the average value of the two chemical shifts seen at the slow exchange limit. In most magnetic resonance shape experiments the isolated resonance lines are not identical, which means that the broadening, coalescence, and narrowing effects do not give symmetrical signals as shown here.*

reaction $A \rightleftharpoons B$, the relationship to the forward and reverse rate constants k_f and k_r are $\tau_A = 1/k_f$ and $\tau_B = 1/k_r$ (cf. Exercise 9-12).

For some special limiting cases, simple formulas for τ_A can be recovered. The simplest case of all is the one shown in Figure 9–2, where two exchanging protons in environments A and B have identical concentrations and identical line widths in the slow exchange limit. Let us define $\tau \equiv \tau_A/2 = \tau_B/2$, let the line separation in the slow exchange limit be Δv_0, and let $(\pi T_2{}^0)^{-1}$ be the line width (full width at half height in sec^{-1}) of each line at the slow exchange limit and of the coalesced single line at the fast exchange limit. (One common situation in which this is true is when the value of $T_2{}^0$ is determined by magnetic field inhomogeneity.) The line shape over the entire broadening, coalescence, and narrowing range for the special case shown in Figure 9-2 is described by the function

$$g(v) = \frac{\frac{1}{4}\tau(v_A - v_B)^2}{[\frac{1}{2}(v_A + v_B) - v]^2 + 4\pi^2\tau^2(v_A - v)^2(v_B - v)^2}$$

We let the line width with exchange, Δv, be given in terms of the transverse relaxation time T_2 by the equation $\Delta v = (\pi T_2)^{-1}$. The following equations hold. For the case in which the two resonance lines barely coalesce (Figure 9-2C)

$$\tau = (\sqrt{2}\,\pi\,\Delta v_0)^{-1}$$

For the initial broadening near the slow exchange limit

$$\tau = \left[\frac{2}{T_2} - \frac{2}{T_2{}^0}\right]^{-1}$$

For the narrowing of a coalesced pair of lines

$$\tau = (\pi\,\Delta v_0)^{-2}\left[\frac{1}{T_2} - \frac{1}{T_2{}^0}\right]$$

Magnetic resonance studies of reaction rates are by no means limited to the simple case we have described. In the first place, the restrictions $[A] = [B]$ and $\tau_A = \tau_B$ can be removed, and numerous additional complications in an ESR or NMR spectrum can be taken into account. The result is either algebraically complicated equations or data analysis by an electronic computer. In either case, it remains quite possible to obtain kinetics data from complex magnetic resonance spectra in the presence of exchange. Another opportunity for rate measurements by magnetic resonance is afforded by special instrumentation. The techniques are called fast passage measurements, multiple resonance, and spin echo measurements. These advanced methods are described in detail in the Supplementary References on magnetic resonance given at the end of the chapter.

The range of reaction rates accessible to investigation by magnetic resonance methods is determined by several characteristics of the particular species being studied. These include the resonance frequencies, chemical shifts, T_2^0 values, nuclear moments, and natural abundances of the nuclei or unpaired electrons in the system under study. For proton NMR experiments, reactions can be studied if the mean lifetimes for exchange are in the approximate range 1 to 10^{-4} sec; for ^{17}O resonance if the mean lifetime range is about 10^{-3} to 10^{-7} sec; for ESR if the mean lifetime range is about 10^{-5} to 10^{-9} sec.

Isomerizations have been extensively studied by NMR line width techniques. Some have been of the hindered internal rotation type, such as

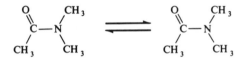

where $k = 2.5 \times 10^8 e^{-11.6/RT}$ sec^{-1}. Another widely studied class of isomerization is ring inversion, such as the chair–chair inversion of monoprotoperdeuterocyclohexane

where $k = 5 \times 10^{13} e^{-11/RT}$ sec^{-1}. Another class of reactions studied by NMR methods is that of proton transfer. An example is methanol

$$CH_3OH + CH_3O^- \rightleftharpoons CH_3O^- + CH_3OH$$

for which $k = 8.8 \times 10^{10}$ sec^{-1} in pure methanol at 298°K.

The principal applications of ESR line width studies of reaction rates have been to electron transfer reactions. An example is

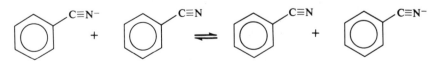

for which $k = 5 \times 10^8$ liters mole^{-1} sec^{-1} was measured at 298°K.

Magnetic resonance line width experiments in the gas phase have been limited to physical rather than chemical interactions of the species studied. This limitation is due principally to the weaker resonance signals that one can obtain from gas samples.

Laser scattering. An entirely different type of line width technique for measurement of reaction rates in solutions may emerge from refined laser technology. A laser beam incident on a solution is predicted theoretically to be scattered by microscopic concentration fluctuations with a line width that depends on, among other things, the reaction rates that restore the fluctuating concentrations to their equilibrium values. Theoretical analyses indicate that first order rate constants in the range 10^2–10^{10} sec^{-1} should be measureable—indeed with unprecedented accuracy. Because of the delicate optical arrangements required for the experiments, laser scattering line width measurements have not been widely applied for measuring reaction rates. It has been demonstrated, however, that the method is indeed feasible in practice. In $ZnSO_4$ solutions, for example, it was found by measuring the line width of scattered 6,328-Å light from a helium–neon laser that some reaction occurs to restore concentration fluctuations on a time scale of 10^{-9} sec.

9-3   RELAXATION SPECTROSCOPY

The subject of relaxation spectroscopy concerns the rate of response of a physical or chemical system to a small disturbance that forces the system to adjust itself to new equilibrium values of its physical or chemical variables. The disturbance may be periodic, for example a sinusoidal change in density produced by a sound wave, or a step function, such as a sudden change in pressure or temperature. The relaxation phenomena that can be studied are related to a great variety of molecular processes. The theory underlying the method, however, is quite general. It requires only that the disturbances be small and that the system always be near equilibrium, in order that the rate equations can be linearized by a method that we discuss later. In the case that several relaxation processes are occurring in the same system, each with a different characteristic response time, called the *relaxation time*, the theory provides an interpretation of the over-all response in terms of the several relaxation times for the individual processes. This separation of relaxation times leads to the term *relaxation spectrum* for the collection of relaxation times in analogy to the optical spectrum that results from separation of optical responses according to frequency.

To see how a relaxation time arises and how it is connected to rate constants, consider the response of the chemical equilibrium

$$A + B \rightleftharpoons C$$

to small perturbations from the equilibrium concentrations $[A]_0$, $[B]_0$, and $[C]_0$. Let the difference between the concentration of C at any time and its equilibrium value $[C]_0$ be the perturbation $x \equiv [C] - [C]_0$; then $[C] = [C]_0 + x$, $[A] = [A]_0 - x$, and $[B] = [B]_0 - x$. The reaction rate is

$$R = \frac{dx}{dt} = k_f ([A]_0 - x)([B]_0 - x) - k_r ([C]_0 + x)$$

Multiplying out the factors in the first term and deleting $0 = k_f [A]_0 [B]_0 - k_r [C]_0$ gives

$$R = -k_f ([A]_0 + [B]_0 - x)x - k_r x$$

By assumption the perturbation x is small compared with $[A]_0$ or $[B]_0$ and may be neglected in the sum, thus giving the *relaxation equation*

$$R = \frac{dx}{dt} = -(k_f ([A]_0 + [B]_0) + k_r)x$$

$$= -\frac{1}{\tau} x$$

with the relaxation time τ given by

$$\tau \equiv (k_f ([A]_0 + [B]_0) + k_r)^{-1}$$

Since k_f and k_r are also related to one another through the equilibrium constant, a measurement of τ gives both k_f and k_r.

Note: The relaxation times τ discussed in this section are related to, but not equal to, the mean lifetimes τ and τ_L of Section 9-2.

The time dependence of x is thus a simple first order exponential decay such as we have encountered frequently before. To derive this, we needed only the condition that the perturbation was small; this is frequently referred to as *linearizing* the differential equation. For other one-step equilibria the derivation of linearized rate equations proceeds in the same way to give appropriate relationships between the relaxation time and the rate constants (Exercise 9-3).

If the perturbation x is introduced as a step function, then the behavior of the relaxation equation is simply exponential decay of the perturbation. We shall discuss several ways of introducing such perturbations and measuring the exponential decays. The original applications of relaxation methods, however, were not to step function perturbations but to sinusoidal perturbations. In the absence of relaxation effects such perturbations are described by differential equations of the form

$$\frac{d^2x}{dt^2} = -\omega^2 x$$

with solutions of the form

$$x = A \sin \omega t$$

where the constant A gives the amplitude of the perturbation and ω is related to the frequency v of the perturbation by $\omega = 2\pi v$. When sinusoidal disturbances are imposed on systems that manifest relaxation effects, then differential equations containing both first and second derivatives of the perturbation x result

$$\frac{d^2x}{dt^2} + \frac{1}{\tau}\frac{dx}{dt} + \omega^2 x = f(t)$$

Such differential equations can be solved, by procedures slightly more advanced than the assumed prerequisite for this book, to yield solutions that for sinusoidal forcing functions $f(t)$ are also sinusoidal, but in which the perturbation is "out of phase" with the imposed disturbance; if the relaxation time is short, the perturbation x reaches its maximum amplitude shortly after the disturbance reaches its maximum amplitude; as the relaxation time becomes larger, the "phase shift" increases. Depending on the nature of the imposed disturbance and the experimental arrangement, these phase shifts can be related to the velocity with which imposed disturbances of a given frequency propagate through the system and also to the rate of damping of the imposed disturbances, that is, to an absorption coefficient, in the system. Relaxation times are then derived by determining the dependence of the propagation velocity or the absorption coefficient on frequency.

For chemical systems in which several elementary reactions have to be considered, the relationships between relaxation times and rate constants have to be derived by a simple matrix algebra that again is just beyond the mathematical level presumed for this book. In essence, there is one relaxation time for each kinetically independent step in the mechanism. If the relaxation times are sufficiently different from one another, then application of relaxation methods can allow determination of the full relaxation spectrum and thence of all of the rate constants.

We turn now to the experimental methods that have been applied to measure relaxation times. The only periodic disturbance method that has proved to be of general utility for studying chemical relaxation is the pressure–

density disturbance associated with propagation of sound waves through liquids or gases. This was first proposed by A. Einstein in 1920 for the $2\,NO_2 \rightleftharpoons N_2O_4$ equilibrium, and subsequently realized for a large number of equilibria as continued improvements in ultrasonic measurement techniques allowed more and more delicate effects to be observed. Ultrasound techniques have contributed little to the study of chemical reactions in gases, although much information about energy transfer in gases has been gathered this way. The major contribution has been rather in the study of ionic equilibria in aqueous electrolyte solutions. Ultrasound methods have the advantages that very fast processes are observable (half-times as short as $\approx 10^{-9}$ sec), that a wide range of relaxation times can be covered ($\approx 10^{-4}$ to $\approx 10^{-9}$ sec), and that many different types of reactions can be studied. They have the disadvantages that large quantities of solution and delicate experimental techniques are demanded. Such reactions as ion solvation and pair formation, proton transfer, and rotational isomerization have been studied with outstanding success.

The most important applications of ultrasound absorption measurements to chemical kinetics have been to the study of electrolyte solutions. The relaxation times found for solutions of various anions and cations are interpreted as interactions between ions—as in ion pair formation—and between ions and solvent molecules. Solutions of salts with divalent anions and divalent cations, for instance, show two ultrasound relaxations. They were found to be attributable to desolvation of each ion during ion pair formation, with the general mechanism being

$$M_{aq}^{2+} + B_{aq}^{2-} \rightleftharpoons (M^{2+} \cdot H_2OB^{2-} \cdot H_2O)_{aq}$$

$$(M^{2+} \cdot H_2OB^{2-} \cdot H_2O)_{aq} \rightleftharpoons (M^{2+} \cdot H_2OB^{2-})_{aq}$$

$$(M^{2+} \cdot H_2OB^{2-})_{aq} \rightleftharpoons (M^{2+}B^{2-})_{aq}$$

The first process is expected to be diffusion controlled, so that the two ultrasound absorptions correspond to establishment of the second and third equilibria. Desolvation of the anion is found to be much faster than desolvation of the cation.

Numerous methods combining a step function disturbance with a fast *in situ* analysis for following the exponential approach to equilibrium have been developed and successfully utilized to investigate chemical relaxation. Once again, the applications have been almost entirely to reactions in solution. The methods are logically classified according to the type of disturbance since the analytical techniques—usually kinetic spectrophotometry or conductivity—do not change as much with the method as with the reaction under study. The most generally useful disturbance is the *temperature jump* since almost all reactions have finite heats of reaction and, therefore, temperature-dependent equilibrium constants. The temperature jump can be effected by passing a sudden electric current through the sample, in which case the heating is by collisions of ions with solvent molecules, or by a microwave pulse, in which case the heating is by rotational energy transfer from polar solvent molecules to the other degrees of freedom of the solution, or by an ir laser pulse, in which case the heating is from transfer of vibrational energy of the solvent molecules to the other degrees of freedom of the solution. In any case the temperature of the entire sample is raised by several degrees (0.1–10 deg) in a time (typically 10^{-6} sec) that is short compared to the chemical relaxation time under study. Relaxation times in the range $1-10^{-6}$ sec can be measured. One special modification of the temperature-jump technique consists of a combination of a stopped flow apparatus with a temperature-jump apparatus. The temperature jump then perturbs the steady state in the flow system. This approach proves to be very effective in eynzyme kinetics experiments.

A second chemically useful disturbance is a *pressure jump*, which is effective for reactions that have substantial volume changes of reaction and, therefore, pressure-dependent equilibrium constants. The pressure jump can be introduced by sudden release of hydrostatic pressure caused by a frangible diaphragm or by a sudden increase of pressure caused by passage of a shock wave. The range of accessible relaxation times is about the same as for the temperature-jump method.

A third disturbance, applicable to the study of equilibria in weak electrolytes, is the *electric field pulse* method. The equilibrium is perturbed in this case by the dissociation field effect, which amounts to an increase in the dissociation constant of a weak electrolyte under the influence of a very large ($\approx 10^5$ V/cm) electric field. The fields are generated by sudden square-wave pulses for ultradilute solutions and by damped harmonic pulses in less dilute solutions in which the temperature rise in the sample can cause perturbations comparable to those caused by the field. Relaxation times as short as 10^{-7} sec are accessible. This method has been applied especially successfully to the study of acid–base equilibria.

The rates of many reactions of hydrogen and hydroxyl ions have been measured by relaxation spectroscopy. Among them are some of the acid–base reactions most familiar to chemistry students:

$$H^+ + CH_3CO_2^- \rightleftharpoons CH_3CO_2H$$

$$H^+ + C_6H_5CO_2^- \rightleftharpoons C_6H_5CO_2H$$

$$H^+ + HCO_2^- \rightleftharpoons HCO_2H$$

$$OH^- + NH_4^+ \rightleftharpoons NH_4OH$$

$$OH^- + CH_3NH_3^+ \rightleftharpoons CH_3NH_3OH$$

Electric field pulse measurements of the rate constants for protonation and deprotonation (i.e., the forward rate constants in the above reactions) show that they are all in the diffusion controlled range, about 10^{10} liters mole^{-1} sec^{-1} (cf. Section 6-5).

Taken as a group, the relaxation methods have two important advantages. First, they are applicable to the study of reactions with an extremely wide range of half-times, resulting in part from technological advances that allow events which proceed at submicrosecond rates to be followed and in part from the simple fact that the reactants are already mixed before the experiment begins. Second, the experiments usually require but small amounts of sample and, once the apparatus has been constructed, small amounts of the experimenter's time. The compensating disadvantages are likewise twofold. First, relaxation spectroscopy is applicable only to reversible (or steady state) reactions. Second, identification of a measured relaxation time with a specific elementary reaction can be quite uncertain unless a considerable amount of supporting evidence is available. This is particularly problematic when a single system shows several relaxation times. Beryllium sulfate solutions, for example, have six relaxation times between 1 and 10^{-9} sec. The three fastest are attributed to hydrolysis, and the three slowest to ion pair formation.

9-4 FLASH PHOTOLYSIS

As in the case of relaxation methods, *flash photolysis* experiments start with a premixed reaction system and perturb it with a sudden pulse, here an intense burst of uv light. In contrast to relaxation methods, in flash photolysis the system may be very far from equilibrium before the experiment starts, and the perturbation is usually very large rather than very small.

A typical apparatus used for flash photolysis experiments is shown in Figure 9-3. The energy discharged in the photolytic flash may range from 1 to 10^3 J. The duration is usually in the range 10^{-6} to 10^{-3} sec, depending on the amount of energy dissipated and the care taken to reduce inductance in the discharge circuit. Ultraviolet light absorbed in the reaction vessel initiates photochemical reactions (Section 10-2). The usual method of measuring reaction rates subsequent to the flash is by taking absorption spectra using less powerful flashes initiated at known delay times after the photolytic flashes. Another method is to monitor the optical transmission through the reaction vessel at a fixed wavelength as a function of time after single photolytic flashes. In this case an experimental record would consist of a photograph of a single trace of an oscilloscope monitoring the transmission. Both methods are called *kinetic spectroscopy.*

Recombination of iodine atoms after flash dissociation of iodine is a classic example of the flash photolysis method. The strong electronic absorption of iodine permits high atom yields in photolysis and accurate

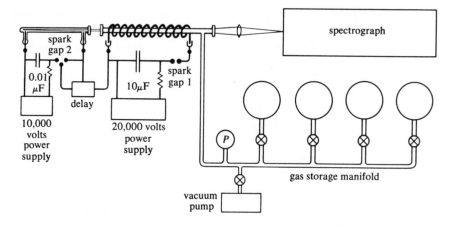

FIGURE 9-3   *A typical flash photolysis apparatus. The helical discharge tube provides intense illumination of the reaction vessel within the helix.   The straight discharge tube provides a shorter, far less intense flash.   It has a continuous emission spectrum in the wavelength region in which the species to be observed have absorptions.   For solution reactions the vacuum manifold would not be required.*

analysis of the recombination progress by kinetic spectroscopy.   In carbon tetrachloride solution the second order rate constant for formation of iodine molecules was found to be 7×10^9 liters mole^{-1} sec^{-1}.   The activation energy was found to be 3 kcal.   These values are in excellent accord with expectations for a diffusion controlled reaction (Section 6-5).   The gas-phase iodine recombination results have already been discussed in Section 5-5.

Additional examples of contributions of flash photolysis to chemical kinetics are mentioned in connection with photochemical mechanisms in Section 10-3.

Flash photolysis has proved to be a fast reaction method of remarkable versatility.   It has contributed to the exploration of many kinds of problems in physical chemistry.   The feature of flash photolysis that makes it so widely useful is that very large concentrations of unstable species can be generated, identified, and studied as a function of time.   Chemical reactions and energy transfer processes have been studied in flash photolysis experiments with gas, liquid, and solid samples.   Numerous elementary reactions of directly identified species have been observed.   In addition, many formerly suspected but undetected species, such as the free radicals CH_2 and NOCN, were discovered and characterized spectroscopically in flash photolysis studies.   A second noteworthy feature of flash photolysis is that irreversible as well as reversible reactions can be studied.

The main limitations of flash photolysis are that suitable absorbing molecules are required and that the fastest rates accessible are limited by the decay time of the photolysis flash, which is generally more than 10^{-5} sec with most conventional flash lamps.   Replacing conventional flash lamps with Q-switched lasers and second-harmonic generators, which have the capability of producing multijoule uv light pulses shorter than 10^{-7} sec in duration, extends the range of accessible reaction rates by about two orders of magnitude.

In the related method of *pulse radiolysis* (Section 10-4) the photolytic flash is replaced by a burst of x rays or high energy electrons.   This method has so far only been applied to radiation chemistry in solution.

9-5   SHOCK TUBES

Another method of studying fast reactions that avoids the mixing problem is provided by the *shock tube*.   A quiescent but potentially reactive gas sample is subjected to adiabatic compression in a shock wave and heated thereby to a temperature high enough that reaction proceeds at a conveniently measurable rate.   A one-dimensional shock wave is generated by confining the propagation to a tube, as in the apparatus shown in Figure 9-4.   In the

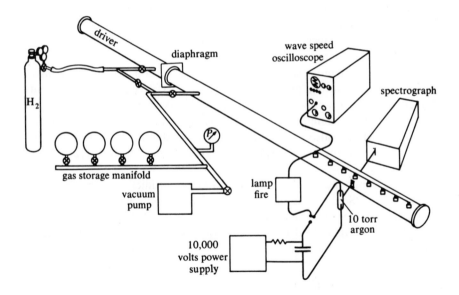

FIGURE 9-4   *A typical shock tube apparatus for chemical kinetics experiments. The driver gas could be hydrogen at a pressure of 2 atm; the experimental gas might be 10 torr of argon containing a few percent of reactants.   The temperature in such a shock wave would be near 2000°K.   In the experiment depicted here, the analytical method is flash spectroscopy, identical to the arrangement shown in Figure 9-2 for a flash photolysis experiment.   A wide variety of analytical methods can be used with shock tubes.*

particular shock tube experiment illustrated, the shock-heated gas flows past an observation station where reaction progress is recorded on an oscilloscope photograph as in kinetic spectroscopy of reactions initiated by flash photolysis. In other arrangements the temperature or the density profile behind the wave can be recorded, or the gas sample can be heated for a short period of time, then quenched in an expansion wave and analyzed later by gas chromatographic or other methods.

The outstanding feature of the shock tube method is the enormous range of pressures and temperatures that is accessible to study. The temperature range 500–5,000°K may be studied with a simple apparatus, and still higher temperatures are attained by more complicated arrangements. Passage of the shock wave through the experimental gas is equivalent to taking it from one thermostat at room temperature to another at, for example, 5,000°K in a time that is typically about 10^{-10} sec. In a simple apparatus such as the one shown, the time resolution of the observations themselves is determined by the traversal time of the shocked gas flowing past the observation station; it usually comes out to be about 10^{-6} sec. In more complicated optical arrangements and especially in experiments at low gas pressures it is possible to follow processes that occur in fewer than 10 encounters.

A second notable feature of the shock tube method is that wall reactions can be disregarded since the observation time is far shorter than the time required for species to diffuse to the walls.

The main disadvantages of the shock tube method are that expensive optical and electronic equipment may be required to carry out the experiments of interest and that it may be necessary to undertake expensive computations with a large digital computer to extract rate constants from the experimental records. There are also gas dynamic limitations that may interfere with the interpretation of the data, especially if exothermic reactions are being investigated.

The shock tube method is applicable to kinetic studies of all categories of gas reactions. Some areas in which significant results have been obtained are studies of the vibrational relaxation and thermal dissociation rates of small molecules, thermal decomposition of large molecules, oxidation reactions, and kinetics of ionization.

9-6 MOLECULAR BEAMS

At very low pressures, the mean free paths of molecules become comparable to the dimensions of the vessel containing them, and they make many collisions with the vessel walls for each collision with other molecules. At pressures below about 10^{-5} torr it is possible to create *molecular beams* that traverse the experimental vessel essentially without undergoing any collisions at all with residual gas molecules. If two such beams are made to cross one another, there will be occasional encounters between molecules from each beam, which results in scattering of beam molecules into directions normally free of them. In molecular beam experiments one examines in the laboratory many of the features of scattering in bimolecular encounters, which was discussed in Section 4-1 as a mere concept helpful in achieving an understanding of the idea of reactive cross section. One apparatus suitable for studying scattering of crossed molecular beams is shown in Figure 9-5.

Crossed molecular beam experiments have been concerned with elastic, inelastic, and reactive channels. In principle it is possible to investigate the dynamics of scattering into each of these channels by studying the intensity of the scattering as a function of the scattering angle and the initial relative velocity of the two beams. The angle dependence can be measured by having a movable detector, and the velocity dependence can be measured by placing velocity selectors in the beams ahead of the scattering point. In addition, there can be velocity selection of the scattered molecules. Simple transformations of the laboratory measurements of scattering angle and scattering intensity to suitable molecular coordinate systems permit conclusions to be drawn about the nature of encounters that lead to scattering into various channels.

In practice the limitations of vacuum and detector technology have so far restricted the application of crossed molecular beam experiments to a few favorable cases. By far the most favorable situations are afforded by ion-molecule reactions and by reactions of alkali metal atoms with halogens, hydrogen halides, alkyl halides, and alkali halides. These reactions have reactive cross sections for attainable relative velocities that are orders of

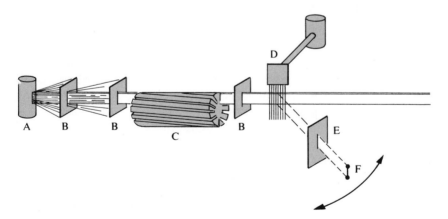

FIGURE 9-5 *Schematic diagram of a simple molecular beam experiment. A— oven for vaporizing alkali metals; B—collimation slits; C—high speed rotor for velocity selection; D—source of second beam; E—detector slit; F—movable detector. The entire set of components is housed in a large evacuated chamber.*

magnitude larger than most other bimolecular reactions. Ions can be detected electrically, and alkali metals and alkali halides happen to be detectable with great sensitivity by their facile ionization on heated metal wires or ribbons. Furthermore, the low vapor pressure of metals and metal halides at room temperature assures that atoms or molecules that are not detected on their flight from the scattering point will condense on the wall of the vacuum chamber and not reach the detector again; similarly, ions are discharged on their first contact with a wall.

Some bimolecular reactions whose reactive cross sections have been studied in crossed molecular beam experiments are

$K + HBr \rightarrow H + KBr$

$K + Br_2 \rightarrow Br + KBr$

$K + CH_3I \rightarrow CH_3 + KI$

$D + H_2 \rightarrow H + HD$

$Cs + RbCl \rightarrow Rb + CsCl$

$Cl + Br_2 \rightarrow Br + BrCl$

$Ar^+ + D_2 \rightarrow D + ArD^+$

It is, of course, unjustifiable to draw general conclusions about bimolecular reactions from the characteristics of alkali metal–halide and ion–molecule scattering. The over-all picture for these reactions is nonetheless very interesting. *Rebound* dynamics, wherein the product molecule (AB in A + BC→AB + C) leaves the encounter backwards, that is, in the general direction from which A approached the encounter, hold for small (about 10 Å^2) reactive cross sections; *stripping* dynamics, wherein AB leaves the encounter in the general direction from which BC approached the encounter, hold for large (about 100 Å^2) reactive cross sections. Intermediate cross sections show intermediate behavior. Some other properties turn out to be independent of the reactive cross section. First, the total elastic, inelastic, and reactive scattering is about the same for reactive encounters as it would be expected to be for the total elastic and inelastic scattering from similar nonreactive molecules. The scattering is *direct*, meaning that the reactive encounters do not last long enough to form a complex that can rotate in space (less than perhaps 5×10^{-13} sec) for scattering of alkali metal atoms from halogens, hydrogen halides, and alkyl halides as well as for scattering of argon and nitrogen ions from deuterium. On the other hand, *indirect* scattering, involving complex formation, does occur in scattering of alkali metal atoms by alkali halides and of HD^+ ions by HD. Finally, it is possible to measure the amount of energy of reaction that appears in relative translational energy of the products departing from the encounter site. The most probable fractions of reaction exothermicity appearing as relative translational energy were found to be 0.05 in $K + Br_2$, 0.4 in $K + CH_3I$, and 0.7 in $Cs + CH_3I$.

The conclusions about reactive scattering drawn from crossed molecular beam experiments are supported by a variety of flash photolysis, fluorescence, and other experiments. Of particular importance are studies of ir fluorescence from the products of $A + BC$ reactions carried out at such low pressure that the observed fluorescence is not perturbed by energy transfer in encounters subsequent to the reactive ones. The observed fluorescence then originates from the quantum states initially produced by reaction, and the distribution of fluorescence intensity is a direct measure of the flux of product molecules into channels corresponding to different product quantum states. Considering the same reactions in the reverse direction, the relative flux information is equivalent to a measure of the reactive cross section as a function of internal quantum states.

Theoretical interpretation of reactive scattering results by the trajectory method (Section 4-4) shows that classical trajectories on plausible potential energy surfaces are capable of reproducing the observations in a qualitative manner. Quantitative comparisons between trajectory calculations and reactive scattering experiments are not very informative, except for $H + H_2$, because the potential energy surfaces themselves cannot be calculated with sufficient accuracy. It is meaningful only to test the trajectory calculations to see if a plausible surface gives the right qualitative description of the scattering. It is then possible to see what qualitative features of the potential energy surface are responsible for various aspects of the scattering. For example, it is found that the fraction of the reaction exothermicity calculated to appear as relative translational energy of the parting products depends markedly on whether the potential energy surface has the reaction energy liberated as the reactants approach one another (an *attractive* surface) or as the products part (a *repulsive*) surface. The relative masses of the atoms are also important in determining the partition of energy between internal excitation and translation.

EXERCISES

9-1 Consider a stirred-flow reactor of volume V cm³ and input feed rates of f_A and f_B cm³ sec⁻¹ of the reactants A and B at concentrations $[A]_0$ and $[B]_0$ moles cm⁻³, respectively. In the output stream $f_{out} = f_A + f_B$ and the concentrations are $[A]_1$ and $[B]_1$. Show that $-d[A]/dt = (f_A [A]_0 - f_{out} [A]_1)V^{-1}$, and explain how the reaction orders in A and B can be found. Assume the reactants are in a large excess of diluent so that the density is constant.

9-2   Reactants A and B are driven at feed rates f_A and f_B cm^3 sec^{-1} into a mixing chamber and thence into a 0.2-cm i.d. tube.   Assume that the mixing is instantaneous and derive expressions for [A] as a function of distance down the tube for (1) second order irreversible kinetics $R = k_2$ [A][B] and (2) first order irreversible kinetics $R = k_1$ [A], with the feed rates f_A and f_B and the initial concentrations [A]$_0$ and [B]$_0$ as parameters.

9-3   Determine the relationships between the relaxation time and the rate constants for the equilibria $A \rightleftharpoons B$ and $2A \rightleftharpoons A_2$ by linearizing the corresponding rate equations.

9-4   The rate constant for $H^+ + OH^- \rightarrow H_2O$ in water at 298°K was found to be 7×10^{13} cm^3 mole^{-1} sec^{-1}.   Calculate the chemical relaxation time of pure water at 298°K assuming $[H^+] = [OH^-] = 1.0 \times 10^{-10}$ moles cm^{-3}.

9-5   It is desired to generate O atoms by flash photolysis of NO_2 with 3,000–3,600 Å light.   In this wavelength region the average quantum yield for dissociation to $NO + O$ (Section 10-3) is 0.95, and the average absorptivity (Section 10-1) is 6×10^3 cm^{-1} torr^{-1}.   Assuming that 5% of the flash energy is converted to light in this wavelength region and that each photon makes a maximum of one traversal of the 1.0-cm diam reaction tube, construct a graph of NO_2 pressure versus flash energy giving 10% decomposition to O atoms.   On the same graph indicate the partial pressure of O atoms provided as a function of flash energy.

9-6   Two important sources of spectral line broadening in addition to Lorentz broadening are *natural* and *Doppler* broadening.   Natural broadening gives a line shape $a(\nu) = a(\nu_0)/[1 + (2(\nu - \nu_0)/\Delta\nu_N)^2]$, of the same form as the Lorentzian line shape, where $\Delta\nu_N = (2\pi\tau)^{-1}$ and τ is the radiative lifetime (Section 10-1).   Doppler broadening gives a line shape $a(\nu) = a(\nu_0) \exp - (2\sqrt{\ln 2} \ (\nu - \nu_0)/\Delta\nu_D)^2$, where the Doppler width $\Delta\nu_D$ depends on the molecular weight M and the temperature according to $\Delta\nu_D = (8 \ln 2RT\nu_0^2/Mc^2)^{\frac{1}{2}}$.   (a) Calculate the temperature at which $\Delta\nu_D = \Delta\nu_N$ for the Lyman-α line of the hydrogen atom, $\lambda = 1{,}216$ Å, $\tau = 1.2 \times 10^{-8}$ sec.   (b) Calculate the foreign gas pressure at which $\Delta\nu_D = \Delta\nu_L$ for the Lyman-α line at 2,000°K, assuming a Lorentz broadening cross section of 200 Å^2.

9-7   Two NMR resonance lines separated by 100 sec^{-1} are found to coalesce to a single line by proton exchange when the temperature is increased to 318°K.   At 298°K, $\Delta\nu_L$ was measured to be 20 sec^{-1}.   Calculate the activation energy of the reaction responsible for the exchange.

9-8   (a) For strong shock waves the shock velocity u is related to the enthalpy increase in the wave ΔH by $u^2 = 2\Delta H M^{-1}$ where M is the molecular weight of the gas.   For constant heat capacity gases $\Delta H = \bar{C}_P \Delta T$.   Calculate the shock velocity required to heat O_2 to 3,000°K assuming (1) translational heating only, $\bar{C}_P = \frac{5}{2}R$; (2) translational and rotational heating, $\bar{C}_P = \frac{7}{2}R$.   To obtain u in cm sec^{-1} ΔH must be in ergs.

(b) Vibrational relaxation in a shock-heated diatomic gas is a slow process, compared with rotational and translational relaxation, that results in cooling of the rotational and translational degrees of freedom.   (1) Calculate the temperature of a shock wave in O_2 moving at the velocity calculated in part (a-1) after vibrational relaxation is complete.   Assume $\bar{C}_P \cong \frac{9}{2}R$.   (2) At $p = 1$ atm the vibrational relaxation time τ_v is given by $\tau_{vp} = 1.5 \times 10^{-10} \exp(133T^{-\frac{1}{3}})$ sec.   Estimate the distance required behind a $p = 1$ atm shock wave in O_2 at the temperature just computed for vibrational relaxation to proceed to the e^{-1} point. The gas flow rate away from the shock front is about $\frac{1}{6}$ of the shock velocity.

9-9   Use the two forms of the Gibbs–Helmholtz equation $d \ln K_{eq}/dp = -\Delta V/RT$ and $d \ln K_{eq}/dT = \Delta H/RT^2$ to estimate the equilibrium constant changes for (1) a pressure jump of 10 atm $\rightarrow$ 1 atm, and (2) a temperature jump of $490°K \rightarrow 500°K$ for the reaction fumaric acid (*trans*-CHCOOH=CHCOOH)$\rightleftharpoons$maleic acid (*cis*-CHCOOH=CHCOOH). Assume that the following pure-compound data are indicative of the behavior to be expected for the solvated species actually under study. Densities: fumaric acid 1.635 g cm^{-3}, maleic acid 1.590 g cm^{-3}. Heats of combustion: fumaric acid 320.0 kcal mole^{-1}, maleic acid 326.1 kcal mole^{-1}.

9-10   A potassium atom beam of 0.015 cm^2 cross-sectional area traverses a 0.75-cm long scattering region containing CH_3I.   (1) What must be the density of CH_3I (molecules cm^{-3}) to attenuate the beam 1 % by reactive scattering if $Q(E) = 35$ Å^2? (2) Calculate the equivalent partial pressure of CH_3I.   (3) If the potassium atom flux is 10^{15} atoms cm^{-2} sec^{-1}, how many KI molecules emerge from the scattering region per second?

9-11   The kinetics of the reaction of O atoms with SO_2 were studied using a stirred reactor and a dual ESR cavity to measure $[O]_0$ and $[O]_1$.   Very little SO_2 was consumed; that is, $[SO_2]_0 = [SO_2]_1$.   The observed decay of $[O]$ was approximately first order.   Suggest a procedure for determining the contributions and the rate constants of the reactions $O + O + wall \rightarrow O_2 + wall$, $O + SO_2 + O_2 \rightarrow SO_3 + O_2$, $O + O_2 + O_2 \rightarrow O_3 + O_2$, $O + SO_2 + O_2 \rightarrow O_3 + SO_2$, and $O + SO_2 + SO_2 \rightarrow SO_3 + SO_2$ to the decay of $[O]$.   Assume that stirred reactor conditions are valid over a wide range of $[O_2]$, $[SO_2]$, $[O]$, and that O atoms are generated by a discharge in O_2 such that $[O]_0 \ll [O_2]_0$.

9-12   The mean lifetime τ_A for species A undergoing a unimolecular reaction $A \rightleftharpoons B$ (cf. Section 9-2) is related to the forward rate constant k^1 by $\tau_A = 1/k^1$; if the reaction is bimolecular $A + B \rightleftharpoons C + D$, the mean lifetime of A is $1/k^2 [B]$.   These formulas are derived by asserting that τ_A is equal to the concentration of A divided by its reaction rate.   Write a qualitative justification for this procedure.   [A mathematical demonstration of the truth of the equation is possible, but not short or easy.]

9-13   Derive relationships between the relaxation times τ (Section 9-3) and the mean lifetimes for exchange τ (Section 9-2) for the reactions $A \rightleftharpoons B$ and $A + B \rightleftharpoons C + D$.

9-14   Derive the normalized line shape function $g(H)$ for ^{17}O nuclear magnetic resonance.   Find the relationship between the mean lifetime τ_L and the line width ΔH_L in the slow exchange limit.

9-15   Microwave absorption spectroscopy can be used to study energy transfer between rotational quantum states and translation or rotation (RT or RR processes). The line width for the $J = 1$ to $J = 2$ rotational transition in OCS was found to be 12.4 $\times$ 10^6 sec^{-1} at 1 torr OCS pressure.   The lifetimes for these two states can be assumed to be approximately equal to one another.   Calculate the average lifetime and the cross section for rotational energy transfer.   Compare your result with the hard sphere collision cross section given in one of the references at the end of Chapter 4.

SUPPLEMENTARY REFERENCES

Fast reaction techniques and results have been described in several general surveys:

E. F. Caldin, *Fast Reactions in Solution* (J. Wiley & Sons, Inc., New York, 1964).

S. Claesson, Ed., *Fast Reactions and Primary Processes in Chemical Kinetics* (Interscience, New York, 1967).

E. M. Eyring, "Fast Reactions in Solutions," in *Survey of Progress in Chemistry*, Vol. II, A. F. Schott, Ed. (Academic Press, New York, 1964), p. 57.

S. L. Friess, E. S. Lewis, and A. Weissberger, Eds., "Investigation of Rates and Mechanisms of Reactions," Chapters XV to XX, in *Technique of Organic Chemistry* (Interscience, New York, 1963), Vol. VIII, Part II.

A description of some of the early advances in the field of fast reaction studies can be found in

The Study of Fast Reactions; Discussions of the Faraday Society **20**, 1956.

The hydrogen atom experiments in flow systems are described in the following references:

F. S. Larkin and B. A. Thrush, *Tenth Symposium (International) on Combustion, Cambridge* (The Combustion Institute, Pittsburgh, 1965), p. 397.

D. J. Le Roy, B. A. Ridley, and K. A. Quickert, *Discussions of the Faraday Society* **44**, 92 (1967).

A. A. Westenberg and N. de Haas, *J. Chem. Phys.* **47**, 1393 (1967).

Theory and experiments on the line shapes in atomic and molecular spectroscopy are presented in

A. C. G. Mitchell and M. W. Zemansky, *Resonance Radiation and Excited Atoms* (Cambridge University Press, Cambridge, 1934), reprinted 1961.

Magnetic resonance experiments on reaction rates are discussed by many authors:

A. Carrington and A. D. McLachlan, *Introduction to Magnetic Resonance* (Harper and Row, New York, 1967), Chapters 11 and 12.

M. W. Hanna, "Electron and Nuclear Magnetic Resonance Spectroscopy," Chapter 9 of *Quantum Mechanics in Chemistry* (W. A. Benjamin, Inc., New York, 1969).

C. S. Johnson, Jr., "Chemical Rate Processes and Magnetic Resonance," in *Advances in Magnetic Resonance*, J. S. Waugh, Ed. (Academic Press, New York, 1965), Vol. I.

J. A. Pople, W. G. Schneider, and H. J. Bernstein, *High-Resolution Nuclear Magnetic Resonance Spectroscopy* (McGraw-Hill, New York, 1959), Chapter 10.

The theory of relaxation spectroscopy is discussed by

G. H. Czerlinski, *Chemical Relaxation* (Marcel Dekker, New York, 1966).

A survey of applications of flash photolysis to chemical problems is given by

G. Porter, "Flash Photolysis and Some of Its Applications," in *Science* **160**, 1299 (1968).

The use of shock tubes for studying chemical reactions is described by

J. N. Bradley, *Shock Waves in Chemistry and Physics* (Methuen and Co., London, 1962).

E. F. Greene and J. P. Toennies, *Chemical Reactions in Shock Waves* (Academic Press, New York, 1964).

Molecular beam experiments are discussed in a number of references:

W. L. Fite and S. Datz, "Chemical Research with Molecular Beams," *Annual Reviews of Physical Chemistry* (Annual Reviews, Inc., Palo Alto, 1963), Vol. 14.

E. F. Greene and J. Ross, "Molecular Beams and a Chemical Reaction," in *Science* **159**, 587 (1968).

J. Ross, Ed., "Molecular Beams," *Advances in Chemical Physics* (Interscience, New York, 1966), Vol. X.

J. P. Toennies, "Molecular Beam Studies of Chemical Reactions," in *Chemische Elementarprozesse*, H. Hartmann, Ed. (Springer Verlag, Berlin, 1968).

Chapter 10

PHOTOCHEMISTRY

AND RADIATION CHEMISTRY

THE ELEMENTARY reactions we have considered so far involved atoms or molecules in their normal, that is, lowest energy or *ground*, electronic states. Reactive encounters were usually distinguished from unreactive ones by their high encounter energy, which was derived from their being on the high energy tail of Maxwell–Boltzmann, or *thermal* distribution functions. Elementary reactions involving ground state atoms or molecules with thermal distributions of encounter energies are often called *thermal reactions* on this account. There is another class of elementary reactions that involves atoms or molecules in *excited*, that is, not ground, electronic states. These reactions are readily studied experimentally. It is also possible to study elementary reactions involving encounters that may not be described by thermal distribution functions.

The simplest way to generate atoms or molecules in excited electronic states is by irradiating the reaction vessel with light at a suitable wavelength. Hence the name *photochemistry* has come to be given to the study of reaction mechanisms in which some elementary reactions involve excited states of atoms or molecules. The most important bimolecular reactions among ground state, but nonthermally distributed, molecules or atoms turn out to be the reactions of atoms that have extremely high translational energy. Unimolecular decompositions of molecules whose vibrational distributions

254

are nonthermal may also be studied. Such elementary reactions, called *hot atom* and *hot molecule* reactions, are treated together with photochemical reactions in this chapter for reasons that will emerge later.

Also included in this chapter is a short survey of the mechanisms of reactions that occur when matter is irradiated with high energy photons (x rays or γ rays) or particles (α rays, β rays, or neutrons).

10-1 ABSORPTION OF LIGHT

If a beam of light of wavelength λ and intensity $I_0(\lambda)$ is directed at a liquid or gas sample which absorbs light at that wavelength, then it is found experimentally that the intensity $I(\lambda)$ of the beam emerging from the sample is given by Beer's law

$$I(\lambda) = I_0(\lambda)\exp(-a(\lambda)[A]x)$$

where $a(\lambda)$ is called the absorptivity, $[A]$ is the molar concentration of the absorbing substance, and x is the path length of the beam through the sample. (This relationship and the coefficient $a(\lambda)$ in it are known by many different names; sometimes concentration units other than moles/liter are used, and sometimes exponentials to base 10 are used.) The intensity can be measured either in $erg\ cm^{-2}\ sec^{-1}$ or in $quanta\ cm^{-2}\ sec^{-1}$ units, with the two being related through the familiar equation

$$\varepsilon = h\nu = hc/\lambda$$

In photochemistry one is more interested in the amount of light energy that is deposited in the sample, which we call the *absorbed intensity* I_a, than in the intensity of the emerging beam I. Obviously, $I_a = I_0 - I$. (When speaking of the intensity of light *absorbed* in a photochemical experiment, the meaning and hence the units of intensity change from erg (or quanta) $cm^{-2}\ sec^{-1}$ for I_0 and I to erg (or quanta) $cm^{-3}\ sec^{-1}$ for I_a, since we are interested in the energy *input* per unit volume into the reaction system rather than in the energy *throughput* per unit area of a beam.) Measuring I_a is not a trivial experimental task. In many photochemical experiments it can be equivalent to measuring the output of the photochemical light source, since I may be much less than I_0. The most reliable photochemical methods for accomplishing an I_a measurement involve *chemical actinometers*. These are photochemical reaction systems whose efficiencies of converting light energy at a given wavelength into some chemical reaction have been accurately measured. One popular actinometer for solution photochemistry is a solution of potassium ferrioxalate, $K_3Fe(C_2O_4)_3$. The photochemical reaction used as the actinometer can be written simply as

$$Fe^{3+} + h\nu \rightarrow Fe^{2+}$$

The oxalate ion is simultaneously oxidized by a complicated mechanism. To use the potassium ferrioxalate actinometer, one exposes the solution to the photochemical light source for a measured length of time and determines, by the 1,10-phenanthroline complex method, the amount of ferrous ion produced. The value of I_a is then calculated using tables that give the efficiency of converting light of various wavelengths into ferrous ion in potassium ferrioxalate solutions.

In irradiated samples of liquids or gases, the light is absorbed by individual molecules (or sometimes atoms). The light energy may have been converted into excitation of the bonding electrons, into molecular vibrations, or into molecular rotations, depending on the identity of the molecules being irradiated and the wavelength of the light. Generally, absorption of wavelengths less than 10^{-4} cm ($= 1\ \mu$) leads to excitation of electrons, or, as is said, *electronic transitions* to *excited electronic states*; absorption at longer wavelengths leads to vibrational or rotational transitions. In photochemistry, electronic transitions are involved. Occasionally, as in photosynthesis for example, the absorbed radiation is in the visible (0.4–$0.8\ \mu =$ 4,000–8,000 Å) region of the spectrum. For most molecules studied by photochemists, wavelengths in the uv region below 4,000 Å are required to cause electronic transitions.

The initial interaction of photochemically active incident radiation with the absorbing molecules thus always results in production of molecules in excited electronic states. In most cases the excited molecules have large amounts of excess vibrational energy in addition to the energy of electronic excitation, that is, the nuclear motions as well as the bonding electrons become energized. Photochemistry involves the manner in which the energy thus deposited in the molecules is dissipated and the mechanisms of the resultant chemical reactions.

Before considering the possible fates of these excited molecules we must introduce some photochemical jargon and a few facts of spectroscopy.

Most molecules have an even number of electrons. In the molecular orbital description of their ground electronic states, these electrons are assigned two at a time to the lowest energy molecular orbitals, the electrons in each orbital being *paired*, that is, each assigned different spin functions designated α and β, or $+\frac{1}{2}$ and $-\frac{1}{2}$. (We ignore here such exceptional cases as CH_2 and O_2, which have ground electronic states in which not all of the electrons are paired.) This situation, in which there are just as many α-spin electrons as β-spin electrons, is called a *singlet state*. Photochemists abbreviate a singlet state with a capital S. The molecular orbital description of the excited electronic states of interest in photochemistry involves one of the electrons being moved from the highest energy orbital that is filled in the singlet ground state S_0 to one of the vacant, still higher orbitals. This can occur with retention of the spin pairing of the ground singlet state, in which case the molecule is excited to one of its *upper*, or *excited* singlet states, designated S_1, S_2, and so forth, in order of increasing energy. Alternatively, the promoted electron can have the same spin function as the electron left in the highest energy orbital of the ground state, in which case the molecule is in one of its *triplet* states designated T_1, T_2, and so on. The first triplet state T_1 is intermediate in energy between the ground state S_0 and the first excited singlet state S_1. The relative energies of higher excited states vary from molecule to molecule. The singlet states are spoken of jointly as the *singlet manifold* since there are many of them, and the triplet states are jointly called the *triplet manifold.*

In addition to naming the electronic states of a molecule, photochemists sometimes need to specify whether the molecule is excited vibrationally. A superscript 0, as $T_1{}^0$, indicates either that a molecule is in the lowest vibrational state or that its vibrational distribution is a thermal one characterized by the temperature of the reaction vessel. A superscript v or m indicates vibrational excitation, as $S_1{}^m$. Figure 10-1 is an energy level diagram illustrating this nomenclature.

Molecular orbital language is also used in photochemistry to describe the electronic transitions themselves. If absorption of a photon results in an electron moving from a bonding π orbital to an antibonding π^* orbital, it is called a $\pi-\pi^*$ transition. Such transitions are characteristic of unsaturated hydrocarbons. Absorption of a photon promoting an electron from a nonbonding n orbital to an antibonding π^* orbital is called an $n-\pi^*$ transition. Carbonyl groups absorb in $n-\pi^*$ transitions.

In order that a photon can be absorbed to excite an electron into a particular higher orbital, the wave functions for the ground state and the excited state must satisfy certain requirements. Some of the requirements have to do with the nuclear positions and motions, and some with the electronic positions, motions, and spins. On this account it is customary to speak of the *vibronic wave functions* of the two states, which we call now for generality the *upper* and *lower* states, or upper and lower *vibronic states* of a given transition. The probability P per unit time that a *radiative transition* will occur between vibronic states ψ'_{ves} and ψ''_{ves} (the single prime referring to the upper state and the double prime referring to the lower state, and the subscripts referring to *v*ibrational, *e*lectronic, and *s*pin) is proportional to, among other things, the square of the transition moment integral

$$P \propto \left\{ \int \psi'_{ves} \hat{M} \psi''_{ves} \, d\tau \right\}^2$$

where $\hat{M}$ is the dipole moment operator. The magnitude of this integral obviously depends on what the wave functions ψ'_{ves} and ψ''_{ves} are. Two aspects of these wave functions are particularly important in photochemistry.

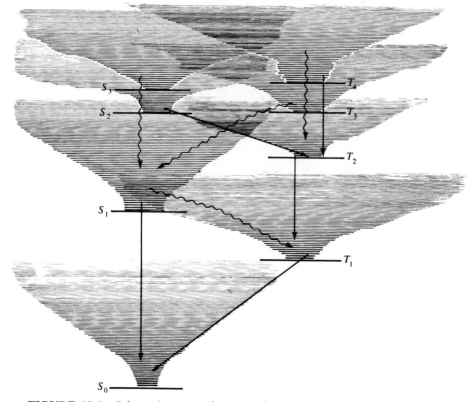

FIGURE 10-1 *Schematic potential energy diagram for a polyatomic molecule, showing the singlet S and triplet T manifolds and some of the possible transitions. The ordinate is the sum of the electronic and vibrational energy of the molecule. Radiative transitions (fluorescence and phosphorescence) are denoted by straight arrows, nonradiative transitions (internal conversion and intersystem crossing) by wiggly arrows. Note that the horizontal lines denoting the vibrational levels of the various electronic states merge at higher energies; this is to indicate that coupling of nuclear and electron motions makes the S and T states become indistinguishable at these energy levels. The actual spacings of the vibrational levels in a typical molecule are much smaller than indicated here.*

First, the integrals become very small if ψ'_{ves} and ψ''_{ves} have different spin multiplicity, that is, if one of them describes an S state and the other a T state. Such transitions are said to be *spin forbidden.* Conversely, transitions in which ψ'_{ves} and ψ''_{ves} both have the same spin multiplicity, either S or T, are said to be *spin allowed*; the spin is said to be *conserved.* Spin conservation is important in determining the magnitude of transition moment integrals in molecules containing low atomic number atoms only; if high atomic number ("heavy") atoms are present, this rule breaks down.

The second important general factor influencing the magnitude of the transition moment integral is called the Franck–Condon principle. It states that only those electronic transitions which can occur without large changes in the positions and the momenta of the nuclei during the transition will have nonvanishing transition moment integrals. In Figure 10-2 this is illustrated for various types of excited electronic states of diatomic molecules. For the extension to polyatomic molecules the internuclear distance coordinate is replaced by a symbolic nuclear geometry coordinate.

The spin conservation rule and the Franck–Condon principle provide the starting points for interpreting the wavelength dependence of the absorptivity $a(\lambda)$, that is, of the *absorption spectrum* of the molecule concerned, since $a(\lambda)$ clearly must be proportional to the transition probability P. The starting

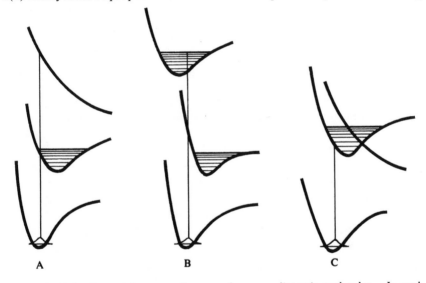

A   B   C

FIGURE 10-2 *Potential energy diagrams for some diatomic molecules. In each case the ordinate is the sum of the electronic and vibrational energy, and the abscissa is internuclear distance. For molecule A, the longer wavelength (lower energy) absorption produces A molecules in the bound electronic state, which gives a molecular band spectrum; the shorter wavelength (higher energy) absorption to the unbound electronic state gives a continuous spectrum. Molecules leave the first excited state by fluorescence or by quenching, and the second electronic state by dissociation. For molecule B the long wavelength absorption is to a bound state, but the consequence of the Franck–Condon principle is that the first excited state produced by the absorption has vibrational energy in excess of the dissociation limit and therefore dissociates immediately, thus giving a continuous absorption spectrum. The second excited state gives a band spectrum. In case C, molecules in the upper state vibrational levels of energy near the energy of the curve crossing dissociate to give ground state atoms. The spectrum is a molecular band spectrum, but perturbed near the transitions corresponding to direct population of the dissociating vibrational levels. Such a curve crossing is called a predissociation.*

point for a photochemical investigation of any molecule is a rationalization of its absorption spectrum in terms of excited vibronic states.

We conclude our brief excursion into the realm of spectroscopy by considering the relationship between emission and absorption of radiation. The dipole moment operator $\hat{M}$ always involves multiplication by some coordinate(s), not differentiation with respect to coordinate(s). This means that $\int \psi'_{ves} \hat{M} \psi''_{ves}\, d\tau = \int \psi''_{ves} \hat{M} \psi'_{ves}\, d\tau$. However, whereas absorption of radiation by an electronic transition can only occur if there is radiation to absorb, emission of radiation, accompanied by demotion of an electron to some lower energy orbital, can occur whether or not the excited molecule is being subjected to radiation from a light beam. The emission rate I_e is first order with respect to the concentration of excited molecules

$$I_e \,(\text{photons sec}^{-1}) = k \,[\text{excited molecules}]$$

The inverse of the first order rate constant for a given transition is called its *radiative lifetime*, denoted by τ. (Note that the radiative lifetime symbol τ has nothing to do with the volume element symbol $d\tau$ used in quantum mechanical integrals.) By the above considerations, the radiative lifetime is inversely proportional to the absorptivity. That is, transitions with large absorptivities (*strong transitions*) have short radiative lifetimes—in the range 10^{-9}–10^{-6} sec—whereas transitions with small absorptivities (*weak transitions*) have longer radiative lifetimes. Spin forbidden transitions can have radiative lifetimes of many seconds.

We now return to photochemistry itself and consider some possible ways that energy added to a molecule by an electronic transition can be dissipated.

10-2   PRIMARY PHOTOCHEMICAL PROCESSES

We shall make a distinction between the elementary steps that the excited molecules undergo within diagrams such as Figures 10-1 and 10-2 and those later elementary steps, which are likely to be thermal reactions, involving dissociation products of the excited molecules or other species generated from them. The former are called *primary*, the latter *secondary* photochemical processes.

In considering the primary processes it is useful to remember that all of the energy deposited in a molecule by absorption of a photon must be accounted for in one way or another. The absorption spectrum will help in many instances by revealing the energies of the various electronic states that may take part. Beyond this, the photochemist must rely on experiments and past knowledge to determine the mechanism by which the energy originally in the photon is eventually distributed in the photochemical system. We now survey the different kinds of routes by which this can occur.

10-2a Vibrational Relaxation

A usual consequence of the Franck–Condon principle is that a large excess of vibrational energy is present in molecules that have just absorbed photons (cf. Figure 9-2). If these molecules are in liquid solution, the excess vibrational energy is very rapidly dissipated to the surrounding solvent molecules, essentially turning directly into heat energy. One expects, therefore, that the first primary process of photochemistry in liquid solution is establishment of a thermal vibrational distribution in the electronically excited molecules. The only exception to this will be some dissociation processes discussed later that occur in times characteristic of single vibrational periods.

If, on the other hand, we are concerned with photochemistry in the gas phase at low pressures, the rate of loss of vibrational energy can be comparable to or slower than the rates of other processes that can occur. Depending on the nature (complexity) of the excited molecules and the molecules, or atoms, with which they undergo encounters, the vibrational energy will be lost with greater or lesser efficiency per encounter, as was discussed already in Chapter 5. The rate of vibrational energy transfer from electronically excited molecules must, therefore, be considered as one of the first problems in sorting out the primary processes in gas-phase photochemistry. An example is given below.

10-2b Fluorescence

The simplest way electronic excitation energy can leave a molecule is through reradiation in the reverse of the electronic transition that put it into the molecule in the first place. This process is named *fluorescence* after the mineral fluorite in which it is prominent. The transition is usually a strong one, with characteristic lifetime in the range 10^{-9}–10^{-6} sec. If the fluorescence is from molecules in the gas phase at low pressure, the average period between encounters can be longer than the radiative lifetime. If the molecules are relatively small ones, the low pressure fluorescence will be from the same vibronic state that results from the irradiation. Such fluorescence is called *resonance fluorescence*. More frequently, one is concerned with liquids or gases at high enough pressure that vibrational and rotational relaxation are faster than fluorescence, in which case the fluorescent emission originates from vibrationally and rotationally relaxed molecules and appears therefore at longer wavelengths than the exciting radiation. For large molecules, resonance fluorescence is not observed since a large molecule can serve as its own heat bath (cf. Exercise 10-12).

A case in which the competition between rotational and/or vibrational relaxation and fluorescence was followed in detail is that of nitric oxide under irradiation by 2,144-Å light from a cadmium vapor discharge lamp. At low pressures ($\sim 10^{-1}$ torr) the only observed fluorescence was from molecules in the $v' = 1$ vibrational state and $K' = 13$ rotational state of the electronic state that results from absorption of the 2,144-Å radiation.   If an inert gas was added, for example N_2 at a pressure of 100 torr, fluorescence was also observed from other rotational states and from the $v' = 0$ vibrational state.   Experiments such as this one, called *fluorescence quenching* experiments, are quite informative about the transfer of vibrational and rotational energy in encounters.   The mechanism may be thought of as a combination of unimolecular radiative processes and bimolecular energy transfer and quenching processes, for example

$$A(S_0) + hv \rightarrow A(S_1^{v=1})$$

$$A(S_1^{v=1}) \rightarrow A(S_0) + hv$$

$$A(S_1^{v=1}) + M \rightarrow A(S_1^{v=0}) + M$$

$$A(S_1^{v=0}) \rightarrow A(S_0) + hv$$

The rates of bimolecular quenching steps are usually expressed in terms of *quenching cross sections* (Section 4-2); they are often large, and usually show small temperature dependence.

The total number of photons emitted as fluorescence is not affected by [M] according to the above mechanism, since those molecules that do not fluoresce from $S_1^{v=1}$ do fluoresce from $S_1^{v=0}$.   In other quenching mechanisms, to be discussed later, the molecules leave the S_1 state entirely and the total fluorescence decreases.

The fraction of absorbed light that is reemitted as fluorescence is called the *quantum yield of fluorescence* and defined by

$$\phi_f \equiv \frac{I_{\text{fluorescence}}(\text{quanta cm}^{-3}\ \text{sec}^{-1})}{I_a(\text{quanta cm}^{-3}\ \text{sec}^{-1})}$$

Anthracene in benzene solution, for example, shows a quantum yield of fluorescence of 0.26 and a radiative lifetime for fluorescence of 4.3×10^{-9} sec.

10-2c   Internal Conversion and Intersystem Crossing

Two nondissociative, radiationless, primary processes compete with fluorescence.   They both involve transitions out of the excited singlet state initially produced.

If the transition is to the triplet manifold, the process, and its inverse, are called *intersystem crossing*. Examples are shown in Figure 10-1. To be meaningful, intersystem crossing to the T_1 state in polyatomic molecules should be understood to include subsequent vibrational relaxation to thermalized T_1 molecules. This is due to the quantum mechanical fact that wave functions describing states of nearly equal energy must be written as linear combinations of the simpler wave functions of, for instance, S_1 and T_1 states alone. Such *mixed* states have both singlet and triplet character in their wave functions, and belong therefore to both the S and the T manifolds. An intersystem crossing from S_1 to T_1 produces a molecule unambiguously belonging to the triplet manifold only after enough energy is lost by vibrational relaxation that the total energy is below the S_1^0 level (Figure 10-1). For many molecules intersystem crossing to the triplet manifold is a relatively efficient process. It is particularly favored in the frequently studied carbonyl compounds.

If the radiationless transition is to another state of the same manifold, such as $S_2 \to S_1^v$, $T_2 \to T_1^v$ or $S_1 \to S_0^v$, it is called *internal conversion*. As indicated in these examples and in Figure 10-1, the internally converted molecule has at first a large excess of vibrational energy. The efficiency of internal conversion is governed by the same considerations as energy transfer in unimolecular reactions, and is therefore greatest for highly excited states with dense vibrational levels. The rate of internal conversion to the ground state, for instance $S_1 \to S_0^v$, is found to be small for most molecules studied by photochemists.

After a molecule has left the electronic state initially produced, it essentially starts all over again to dissipate energy in its new electronic state.

10-2d Phosphorescence

Although the transition moment integrals for radiative S–T and T–S transitions are small, they are not zero, and emission from T–S transitions, called *phosphorescence*, is in fact sometimes observed. As with fluorescence, it is customary to state the rate by a quantum yield of phosphorescence ϕ_p, defined by

$$\phi_p \equiv \frac{I_{\text{phosphorescence}}(\text{quanta cm}^{-3} \text{ sec}^{-1})}{I_a(\text{quanta cm}^{-3} \text{ sec}^{-1})}$$

Quantum yields for phosphorescence are usually small. Notable exceptions are substituted aromatic compounds at low temperature, and in the gas phase the diketone biacetyl, CH_3—CO—CO—CH_3. Since the transitions are spin forbidden, the radiative lifetimes are much longer than for S–S or T–T transitions, ranging from 10^{-3} to 1 sec for typical cases.

10-2e   Excitation Transfer

Electronic energy stored in one molecule can be transferred in a bimolecular reaction, which can be considered a special type of fluorescence or phosphorescence quenching, to a molecule of an entirely different compound. The molecule releasing the energy is then called the *donor* molecule, and the one receiving is the *acceptor* molecule. The first requirement for this process to occur is that the donor must have sufficient excitation energy to cause excitation to an excited state of the acceptor; the nearer the two excitation energies are to one another the greater the transfer probability will be. The second requirement is fulfillment of certain rules regarding the spin states and the symmetries of the donor and acceptor wave functions. It is found that favored S–S transfers can occur over large (~ 100 Å) distances; that is, large reaction distances must be assumed in order to match the measured transfer rates with the equations for diffusion controlled reactions. The corresponding distances for S–T transfer turn out to be about the same as the molecular dimensions.

The rates of excitation transfer reactions vary widely for a given donor, depending on the spectrum of electronic states of different acceptors. For T–T transfer from biacetyl (whose T_1 level lies 56 kcal above the S_0 level) in benzene solution, for example, the rate constants were found to range from 2×10^3 liters mole^{-1} sec^{-1} for phenanthrene as acceptor (T_1–$S_0 = 62$ kcal) to 8×10^9 liters mole^{-1} sec^{-1} for anthracene as acceptor (T_1–$S_0 = 42$ kcal). Experiments on other donor–acceptor combinations confirm the generalization that these transfers are favored by large positive values for the quantity $(T_1$–$S_0)_{donor} - (T_1$–$S_0)_{acceptor}$.

A special case of excitation transfer is for two T_1 triplets of the same molecular species to combine as donor and acceptor to form one S_1 and one S_0 molecule. The phenomenon is known as *triplet–triplet annihilation*. It is clearly favored by large triplet concentrations, such as can be formed in flash photolysis experiments. The S_1 fluorescence arising through this indirect route will decay much more slowly than the S_1 radiative lifetime would indicate, for which reason the term *delayed fluorescence* is used to describe it.

Another case of excitation transfer is *self-quenching* of S_1 molecules in solution by S_0 molecules of the same kind. The unusual feature of this bimolecular route of energy dissipation is that it may proceed by way of an intermediate complex, called an *excimer*, which may itself fluoresce.

A final, important case of excitation transfer is *chemical quenching*, whereby the acceptor molecule is excited into an unbound, or weakly bound, electronic state that dissociates immediately into free radicals or atoms. This effect can be used to induce *photosensitized reactions*. For example, quenching Hg atoms from the T_2 excited state to the S_0 state can give the elementary reactions

$$Hg\ (T_2) + H_2 \rightarrow Hg\ (S_0) + H + H$$

$$Hg\ (T_2) + CH_4 \rightarrow Hg\ (S_0) + CH_3 + H$$

$$Hg\ (T_2) + N_2O \rightarrow Hg\ (S_0) + N_2 + O$$

Since excitation of Hg to the T_2 state is experimentally simple, namely, by irradiation with a mercury vapor lamp, these photosensitized reactions are convenient routes to reactive intermediate species such as H or O atoms or CH_3 radicals.

10-2f Isomerization

Irradiation of a molecule containing an olefinic double bond can result eventually in an electronic state in which rotation about the originally double bond is possible. In this case *cis–trans* isomerization will occur as a primary photochemical process. Some photochemical isomerizations lead to startling molecular rearrangements, for example,

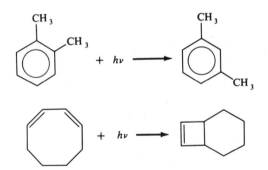

The mechanism(s) by which isomerizations occur is complicated and so far only partially clarified.

10-2g Dissociation

The last possible primary process we consider is certainly the most important one for chemistry. Photolytic rupture of chemical bonds can occur in several ways, some of which are illustrated in Figure 10-2. One simple possibility is direct excitation to an unbound, or *repulsive*, electronic state, forming two fragments on a time scale comparable to molecular vibrations, $\sim 10^{-13}$ sec. The other possibilities are repulsive states, or else bound states (with binding energy less than the available vibrational energy) that are isoenergetic with one or more of the bound electronic states through which a molecule can pass during its progressive energy degradation. With the isoenergetic repulsive state, the time scale for dissociation is the same as the time scale for collisional energy transfer. With dissociation by way of an isoenergetic bound state, the time scale for dissociation can combine the time scales for collisional energy transfer and for unimolecular decomposition. Dissociation occurs from the S manifold, the T manifold, or both.

In diatomic molecules the only variations in dissociation products will be with regard to the electronic states of the atoms produced and the relative translational velocities of the parting atoms. For the direct photodissociation route these variations are readily interpretable from the absorption spectrum. In small polyatomic molecules the absorption spectrum also provides important information about dissociation pathways. For example, ozone, which has absorption bands both in the visible and uv regions, is photodissociated by way of the two paths

$$O_3 \ (S_0) + h\nu \ (\text{red}) \rightarrow O_2 \ (T_0) + O \ (T_0)$$

$$O_3 \ (S_0) + h\nu \ (\text{uv}) \rightarrow O_2 \ (T_0) + O \ (S_1)$$

For the indirect pathways, kinetics experiments are required to find the mechanism. An example is the dissociation of I_2 by green (5,461 Å) light, which proceeds by way of a repulsive T_1 state that at one internuclear distance is isoenergetic with a vibrational level of the T_2 state

$$I_2 \ (S_0) + h\nu \rightarrow I_2 \ (T_2'')$$

$$I_2 \ (T_2'') + M \rightarrow I_2 \ (T_2''') + M$$

$$I_2 \ (T_2''') + M \rightarrow I_2 \ (T_1) + M$$

$$I_2 \ (T_1) \rightarrow 2I$$

Here the initial excitation is directly to the T_2 state, an exception to the usual spin conservation rule caused by the strong "spin–orbit coupling" which occurs when atoms with high atomic number are present. Collisional energy transfer lowers the vibrational energy until the T_2 state is isoenergetic and at the same internuclear separation (Figure 10-2) as the repulsive T_1 state, at which point internal conversion and dissociation occur.

A great variety of primary photochemical dissociations can occur in polyatomic molecules. We list here only a few illustrative examples of the many photodissociations that have been discovered.

$$CH_2N_2 \; (S_0) + h\nu \rightarrow CH_2 \; (T_0) + N_2$$

$$\rightarrow CH_2 \; (S_1) + N_2$$

$$CH_3COCH_3 + h\nu \rightarrow CH_3CO + CH_3$$

$$H_2O_2 + h\nu \rightarrow OH + OH$$

$$CH_3Br + h\nu \rightarrow CH_3 + Br$$

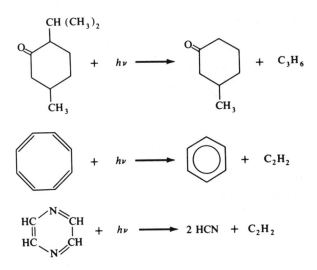

The inverse to photodissociation can also be observed. It is one example of *chemiluminescence*, light emission resulting from chemical reaction. In the case of two atoms, the inverse to photodissociation is called *radiative recombination*. An example of a chemiluminescence mechanism is

$$O \; (T_0) + CO \; (S_0) + M \rightarrow CO_2 \; (T_1) + M$$

$$CO_2 \; (T_1) + M \rightarrow CO_2 \; (S_1) + M$$

$$CO_2 \; (S_1) \rightarrow CO_2 \; (S_0^v) + h\nu$$

The light produced in this sequence gives a characteristic blue glow to carbon monoxide flames.

10-2h Résumé

Photon energy originally deposited in absorbing molecules as electronic excitation can have a variety of fates, depending on the nature of the absorber, the wavelength of the light, and the experimental conditions. This variety is both complicating and convenient for the photochemist since the number of processes to be studied and the number of possible experiments increase together. One guiding principle should not be forgotten in considering the primary photochemical processes a molecule can undergo: the sum of the quantum yields (on an energy basis) must be unity, since energy must be conserved in the system.

10-3 SECONDARY PHOTOCHEMICAL PROCESSES

Dissociation of excited molecules may provide reactive intermediates that can undergo *secondary* elementary reactions of a thermal nature. The mechanisms composed of these reactions are the second major object of research in photochemistry. In principle, there are no differences between the thermal elementary reactions in which photochemically produced intermediates can take part and the elementary reactions in which the same intermediates otherwise produced can take part. In practice, however, only a limited number of ways have been discovered to gather information on the elementary reactions of any given intermediate species, and as a consequence there are many elementary reactions and many mechanisms that have been studied primarily or exclusively by photochemical means.

The efficiencies of secondary photochemical processes are expressed as *quantum yields* ϕ_i defined as

$$\phi_i \equiv \frac{\text{Rate of production of substance } i \text{ (molecules cm}^{-3} \text{ sec}^{-1})}{I_a \text{ (quanta cm}^{-3} \text{ sec}^{-1})}$$

Whereas quantum yields for fluorescence and phosphorescence have maximum values of unity, chain reactions allow product quantum yields to have any values (cf. Section 8-4).

In this section a sampling of photochemical mechanisms is given in order to convey a feeling for what can be discovered about mechanisms per se by photochemical experiments. The mechanisms in each case are those suggested by the experimenters and rationalize only the main reaction pathways.

Photoreduction of benzophenone in solution

$$\phi_2 CO\ (S_0) + hv \rightarrow \phi_2 CO\ (S_1)\ n-\pi^*$$

$$\phi_2 CO\ (S_1) \rightarrow \phi_2 CO\ (T_1)$$

$$\phi_2 CO\ (T_1) + \text{solvent} \rightarrow \phi_2 \dot{C}OH + \text{solvent radical}$$

$$2\phi_2 \dot{C}OH \rightarrow \phi_2 COHCOH\phi_2$$

Photochemical reduction of aromatic ketones was discovered at the turn of the century as an effect of sunlight. The T_1 state was identified as the H-atom abstracting species through a variety of experiments, including the absorption spectrum after flash photolysis. In some solvents the photo-reduction quantum yield is found to be unity.

Photosensitized oxidation of anthracene in oxygenated solution

Mechanism I:

$$\text{dye}\ (S_0) + hv \rightarrow \text{dyebiradical}$$

$$\text{dyebiradical} + O_2 \rightarrow \text{dyebiradical} \cdot O_2\ \text{complex}$$

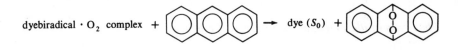

Mechanism II:

$$\text{dye}\ (S_0) + hv \rightarrow \text{dye}\ (S_1)$$

$$\text{dye}\ (S_1) + O_2\ (T_0) \rightarrow \text{dye}\ (S_0) + O_2\ (S_1)$$

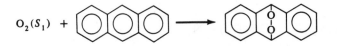

Photochemical experiments alone do not allow a decision between the two possible mechanisms. Comparison experiments in which $O_2(S_1)$ is generated chemically, however, show that the second mechanism is correct.

Ultraviolet photolysis of ozone in the gas phase

$$O_3\ (S_0) + h\nu(2{,}000\text{--}3{,}100\ \text{Å}) \to O_2\ (T_0) + O\ (S_1)$$

$$O\ (S_1) + O_3\ (S_0) \to O_2\ (T_0^{\ 0}) + O_2\ (T_0^{\ v})$$

$$O_2\ (T_0^{\ v}) + O_3\ (S_0) \to 2\ O_2\ (T_0) + O\ (S_1)$$

$$\to 2\ O_2\ (T_0) + O\ (T_0)$$

$$O\ (S_1) + M \to O\ (T_0) + M$$

$$O_2\ (T_0) + O\ (T_0) + M \to O_3\ (S_0) + M$$

The presence of a chain reaction was indicated by classical photochemical experiments that gave quantum yields of O_3 disappearance that were greater than unity. Flash photolysis experiments permitted detection of the vibrationally excited O_2 intermediate and thus supported the complete chain mechanism shown. This set of reactions is part of a photochemical steady state that determines the chemistry of the earth's upper atmosphere. The presence of ozone prevents sunlight in the wavelength range 2,000 to 3,100 Å from reaching the earth's surface. These reactions also combine with others involving air pollutants to produce photochemical smog, which is an important subject of current photochemical research. During the evolution of the earth's atmosphere, other photochemical steady states played important roles.

Photolysis of diazomethane in presence of ethylene

$$CH_2N_2\ (S_0) + h\nu \to CH_2N_2\ (S_1^{\ v})$$

$$CH_2N_2\ (S_1^{\ v}) + M \to CH_2N_2\ (T_1^{\ v})$$

$$CH_2N_2\ (S_1) \to CH_2\ (S_1) + N_2$$

$$CH_2\ (S_1) + M \to CH_2\ (T_0) + M$$

$$CH_2N_2\ (T_1) \to CH_2\ (T_0) + N_2$$

$$CH_2\ (S_1, T_0) + CH_2{=}CH_2 \to C_3H_6^*$$

$$C_3H_6^* + M \to \text{propylene and cyclopropane} + M$$

The methylene diradical CH_2 undergoes two characteristic reactions: *insertion* into a C—H single bond and *addition* to a C=C double bond. Distinction between the reactions of T_0 and S_1 methylene radicals is a difficult matter due to the coupling between the electronic and vibrational motions in CH_2 (the electronic wave function changes from S to T character as the bond angle changes from that characteristic of the parent CH_2N_2 molecule to the linear T_0 bond angle of CH_2) and to the fact that the diluent M provides energy transfer for all of the intermediate species present.

Photolysis of 2-pentanone

$$CH_3CH_2CH_2COCH_3 \; (S_0) + h\nu \rightarrow CH_3CH_2CH_2COCH_3 \; (S_1)n - \pi^*$$

$$CH_3CH_2CH_2COCH_3 \; (S_1) + M \rightarrow CH_3CH_2CH_2COCH_3 \; (T_1) + M$$

$$CH_3CH_2CH_2COCH_3 \; (S_1 \text{ or } T_1) \rightarrow C_2H_4 + CH_3COCH_3$$

Product formation occurs through a six-membered ring that allows H-atom transfer from the γ carbon to the oxygen atom, thus giving the enol form of acetone. Showing that both the S_1 and T_1 states undergo the reaction was accomplished by independent generation and removal of the T_1 molecules. The essential requirement for reaction appears to be having sufficient vibrational energy in the molecule to form the transition state ring.

A characteristic of photochemical mechanisms that is not fully expressed in these examples is that of complexity: there are usually many subsidiary products to be found in photolyzed reaction mixtures and many alternative pathways that can account for their formation. In photochemistry more than any other branch of kinetics it is important to exercise special care in purification of reagents, diligence in analyzing for all of the minor reaction products, and a judicious blend of imagination and circumspection in rationalizing the generally copious results.

10-4 RADIATION CHEMISTRY

Radiation chemistry deals with the mechanisms of chemical reactions that occur upon absorption of *ionizing radiation*, which is an inclusive term for high energy (e.g., more than 50 eV) charged particles, such as α rays, β rays, cosmic rays, or particles from an accelerator, and for ultrashort wavelength electromagnetic radiation (e.g., $\lambda < 10$ Å), which is classified as γ rays if it is nuclear in origin and x rays if extranuclear in origin. There are several strong similarities to photochemistry, for example in the procedures of irradiation, manner of expressing conversion efficiency, and differentiation between primary and secondary processes. On the other hand, there is one important difference between radiation chemistry and photochemistry. In an idealized photochemical experiment all of the absorbed radiation appears photon for photon in single molecules excited to some specific electronic state. In a radiation chemistry experiment, the absorbed energy is deposited without selectivity in whatever molecules are in the path of the radiation, thus causing ionization and excitation to all sorts of electronic states. These ions and excited species start out in high concentration along the *track* of each incoming particle or ray and can react with one another or with unexcited molecules to give a bewildering complexity of products. The clearest distinction between radiation chemistry and photochemistry would be for reactions in dilute solution: in photochemistry the radiation is virtually all absorbed by the solute, but in radiation chemistry by the solvent.

The simplest measure of the amount of ionizing radiation being absorbed in a gas sample is obtained by collecting the positive and negative ions on electrodes and measuring the ion current. This led to the use of the *ion pair yield*, defined as the ratio of the number of molecules of a given substance formed or destroyed per ion pair formed, as the first radiation chemistry equivalent of the quantum yield in photochemistry. Unfortunately, it was not possible until quite recently to measure the rate of ion pair formation in liquid or solid samples. Therefore, the ion pair yield was eventually supplanted by the *G value* as the measure of radiochemical yield. The *G* value is defined as the number of molecules of a given substance formed or destroyed in a given system per 100 eV of a given type of ionizing radiation energy absorbed. If the average amount of energy W deposited per ion pair formed happens to be known for the particular system being irradiated (typically 30–40 eV; 32.5 V in air), then the two measures can be related to one another.

As in photochemistry, measuring the quantity of energy absorbed is the first requirement for quantitative experimentation. The actinometer of photochemistry becomes the *dosimeter* of radiation chemistry. For x rays or γ rays, dosimetry is usually accomplished by measuring the rate of ion production in a known volume of air exposed to the radiation source. For charged particles, the radiation is simply collected and measured as a current on a sensitive ammeter. Perhaps more convenient than either of these electrical methods are *chemical dosimeters*, the most common of which involves $Fe^{2+} \rightarrow Fe^{3+}$ oxidation in a ferrous sulfate solution.

The primary processes of radiation chemistry vary more with the type and energy of the radiation than with the nature of the absorber, and on this account it is common to discuss them primarily as radiation energy dissipation mechanisms.

Electrons lose their energy mainly by converting it to x radiation, called bremsstrahlung, when the incident electrons approach absorber nuclei, and by converting it to translational energy in collisions with absorber electrons. *Heavy charged particles*, such as protons, α particles, or fission fragments, lose energy in collisions with absorber electrons. *Neutrons* interact only with nuclei and produce ionization indirectly through the charged nuclei that recoil at high velocity from elastic collisions with the neutrons, and also through the γ rays or protons that result from neutron–nuclei reactions. In all of the particle–matter interactions, energy is lost piecemeal in a large number of small energy transfers. For high energy photon–matter interactions the opposite holds, and energy is lost in a few large steps. High energy (more than about 10 MeV) γ *rays* can interact directly with nuclei in atomic equivalents of molecular photochemical primary processes. Photons with energies greater than 1 MeV can also disappear into creation of an electron–positron pair; the positron later recombines with an electron yielding two 0.51 MeV γ rays. Lower energy *x-ray* photons can disappear by the photoelectric effect in ejection of electrons from the inner cores of absorber atoms, or can undergo Compton effect encounters with valence electrons.

Many of the energy dissipating steps create photons or charged species that are themselves ionizing radiation and undergo their own energy dissipation processes. Secondary electrons with energies greater than 100 eV are produced in about half of the ionizations produced by a primary electron, proton, or α particle. These secondary electrons, known as δ *rays*, branch away from the track of the primary particle. The secondary electrons with energy less than 100 eV have a short range in liquids or solids, and a *spur* of ionization with perhaps 2–3 ion pairs and twice as many electronically excited molecules appears on the track of the primary particle.

The net result of the energy dissipation is production of ions and electronically excited atoms and molecules along the path of the radiation. It is found that the nature of the radiation does not influence the identity of species formed in a given absorber to an appreciable degree, but it does influence their local concentrations if the absorber is a solid or a liquid. The different chemical effects of different types of radiation in a given system are due to the different densities of reactive species produced along the track of the individual particles. These densities are described by the *linear energy transfer* (LET) of a given type of radiation in a given absorber, which is defined simply as the derivative dE/dx, where E is the energy of the radiation and x is the path length in the absorber, and expressed in keV per micron. Typical mean values in water range from 1 for x rays to 100 for α particles in these units. For radiation of equal energy, LET values increase in the order: γ rays, β rays, α rays, fission fragments, that is, in order of increasing mass. This wide LET range implies a corresponding wide range in the density of spurs and single ionizations and excitations along the tracks of different types of radiation. For α particles the spurs are so close together that they form a continuous column of ionization and excitation, whereas for γ rays in water the spurs can be 10^{-4} cm apart on the average. Therefore, the relative contributions of reactions involving two primary species can be quite different for different types of radiation.

The primary processes of radiation chemistry are studied more by physicists than by chemists; physicists leave the secondary thermal reactions of the primary products to be the main concern of radiation chemists.   In liquids and solids there are rapid sesquiary processes that are interposed between the primary radiation–matter interaction and the secondary thermal reactions of the ions and excited molecules.   These include solvation of ions and electrons, dissociations, intersystem crossings and internal conversions, and, in general, thermalization of all species along the track such that the secondary reactions are mostly thermal reactions of ions, T_1 molecules, and free radicals.   To enter into a survey of the mechanisms of radiochemistry is impossible here because of the variety of primary products and the variety of secondary reactions these reactive species undergo.   We must be content with a single example.

A "simple" radiation chemistry mechanism would be that for irradiation of oxygen.   Denoting primary processes by —$\wedge\wedge$→, electronic or vibrational excitation indiscriminately by *, and possible excitation by (*), a postulated mechanism, based on a wide variety of observations, is

$$O_2 \xrightarrow{\wedge\wedge} O_2^+ + e^-$$

$$O_2 \xrightarrow{\wedge\wedge} O_2^*$$

$$O_2 + e^- \rightarrow O_2^-$$

$$O_2^+ + e^- \rightarrow O + O(*)$$

$$O_2^+ + O_2^- \rightarrow O + O_3$$

$$O_2^+ + O_2^- \rightarrow O_2^* + O_2$$

$$O_2^+ + O_2^- \rightarrow O_2 + O(*) + O(*)$$

$$O + O_2 + M \rightarrow O_3 + M$$

$$O_2^* \rightarrow 2O$$

$$O_2^* + O_2 \rightarrow O + O_3$$

$$O + O_3 \rightarrow O_2 + O_2(*)$$

$$O + O_3 \rightarrow O_2 + O + O$$

The investigation of radiation chemistry mechanisms in aqueous solution is a subject of greater complexity than this, and clearly of far greater importance in connection with radiation biochemistry.

We conclude this brief introduction with allusions to two closely related areas of chemical kinetics. The first is the chemistry of electric discharges, in which similar primary and secondary processes are found. The second is to the study of bimolecular ion–molecule reactions in mass spectrometers with inlet systems modified to tolerate high enough inlet pressures for such reactions to occur, and to the study of unimolecular ion decomposition reactions during the free flight of ions in normal, or modified, mass spectrometers. In fact, mass spectrometer experiments provide not only independent confirmation of the existence of elementary ion–molecule and ion decomposition reactions postulated to rationalize the results of conventional radiation chemistry experiments, but also knowledge of the rate constants of these reactions. In favorable cases one can measure the dependence of the bimolecular rate constants on relative energy of translation and the dependence of the unimolecular rate constants on the degree of energization by electron impact.

10-5 HOT ATOM AND HOT MOLECULE REACTIONS

In Section 9-6 the use and limitations of energy-selected molecular beams for measuring the reactive cross sections of bimolecular reactions were described. Another kind of experiment providing similar information is *hot atom* chemistry. Atoms with velocities far in excess of average thermal (10^5 cm sec^{-1}) velocities can be generated in a few favorable cases by photochemical methods or by nuclear reactions. With delicate analytical methods the products resulting from reactions of these hot atoms can be determined. Nonthermal unimolecular reactions are, in the broad sense, observed in every photochemical experiment; if we restrict the sense of *hot molecule* chemistry to mean the special reactions of molecules with exceptionally large amounts of vibrational energy, then we have a class of reactions that will provide detailed information about the nature of unimolecular decomposition or isomerization rather analogous to cross-section information about bimolecular reactions derived from molecular beam or hot atom experiments.

Hot atom generation is a consequence of the conservation of momentum in photochemical decompositions and nuclear reactions. To take an example of photochemical decomposition first, uv photolysis of HI results in bond rupture to form H atoms and I atoms in either the ground or the first excited atomic state

$$HI + h\nu \rightarrow H + I$$

$$\rightarrow H + I^*$$

The energy of the absorbed photon is far greater than the bond energy of the HI molecule, and the excess energy must be divided between the parting H and I atoms in such a way that their combined momentum is equal to the original momentum of the HI molecule. If the HI molecule was originally at rest, for example, then the H and the I atoms leave in opposite directions such that

$$m_H v_H + m_I v_I = 0$$

Rearranging and squaring gives

$$m_H{}^2 v_H{}^2 = m_I{}^2 v_I{}^2$$

Solving for the kinetic energy of the H atom

$$\frac{m_H v_H{}^2}{2} = \varepsilon_H = \frac{m_I}{m_H} \varepsilon_I$$

Since the atomic weight ratio m_I/m_H is 127, more than 99% of the excess energy appears as translational motion of the H atom. If 2537-Å light from a mercury lamp is used for photolyzing HI, the difference between the photon energy, $hc/\lambda = 0.395 \times 10^5$ cm^{-1} $\times$ 2.86 cal cm = 113 kcal, and the H—I bond dissociation energy, 71 kcal, is 42 kcal, more than 99% of which is H-atom translational energy. Since 42 kcal is large compared to the activation energies of many bimolecular exchange reactions, H atoms created by 2,537-Å photolysis of HI can be very reactive.

Atoms with still higher energies are generated in nuclear reactions. Tritium atoms, for instance, are created with energies of megacalories by thermal neutron irradiation of ^{3}He and ^{6}Li

$$^3He + n \rightarrow p + {}^3H \ (0.19 \text{ MeV} = 4.4 \times 10^6 \text{ kcal})$$

$$^6Li + n \rightarrow \alpha + {}^3H \ (2.7 \text{ MeV} = 6.2 \times 10^7 \text{ kcal})$$

As their velocity is reduced by collisions, the ^{3}H atoms enter into the energy range of chemical reaction (tens or at the most hundreds of kilocalories) from above and undergo characteristic bimolecular reactions. The radioactivity of the ^{3}H atoms allows their locations in product molecules to be found with precision and sensitivity. It is found, for example, that the thermal reaction of abstraction

$$^3H + CH_4 \rightarrow CH_3 + H^3H$$

is accompanied for the hot ^3H atoms by the alternatives

$$^3H + CH_4 \rightarrow CH_3\,^3H + H$$

$$^3H + CH_4 \rightarrow CH_2\,^3H + H_2$$

Molecules with large amounts of vibrational energy are part of the general energy degradation mechanism describing the photochemistry of almost every polyatomic molecule, but only in special cases is it possible to derive information about the relationship between the degree of vibrational excitation and unimolecular reactivity. The most successful experiments have involved hot free radicals generated by some special recombination reaction. In radiation chemistry there are suitable neutralization reactions

$$A^+ + e^- \rightarrow R\cdot^* + R'\cdot$$

The hot radical $R\cdot^*$ then undergoes decomposition or other reactions that are not found for thermal $R\cdot$. A simple but informative gas-phase hot radical generation reaction is the addition of H atoms to olefins

$$H + RR'C{=}CHR'' \rightarrow RR'\dot{C}CH_2R''^*$$

yielding a hot radical that can have numerous exothermic decomposition and isomerization routes that can be studied as functions of the different R, R', and R'' and the pressure and nature of a coolant gas introduced to promote thermalization of the hot radical in competition with its decomposition. Bimolecular reactions of hot radicals are, of course, also possible and sometimes postulated.

The relative reactivities of different vibrational states of diatomic molecules, mentioned in Section 4-1, would form still another class of nonthermal reactions.

EXERCISES

10-1 Construct a log–log graph of the energy in kcal associated with light in the wavelength region 1,000–10,000 Å. Indicate on the energy scale the bond dissociation energies for O_2—O, I—I, HO—OH, Cl—Cl, CH_3—CH_3, H_2, HO—H, N—O, and C—O.

10-2 The radiative lifetime of Hg atoms for T_2–S_0 phosphorescence is 1.1×10^{-7} sec. The quenching rate constant for encounters with H_2 is 2.9×10^{11} liters mole^{-1} sec^{-1} at 25°C. (a) Calculate the quenching cross section. (b) What pressure of H_2 is required to reduce the phosphorescence intensity by a factor of 10?

10-3   The Fe^{2+} quantum yield for the ferrioxalate actinometer is 1.24 at 3,130 Å. Calculate the emission intensity (watts) at this wavelength required to generate 10^{-6} moles Fe^{2+} in a 10-sec exposure, assuming that all of the radiation is absorbed by the actinometer.

10-4   Photolysis of CH_2CO at 3,130 Å produces CO with quantum yield of 2.0 and C_2H_4 with quantum yield of 1.0. How much CO would be formed in a 1-h irradiation with the lamp described in the previous problem in a cell in which all of the incident 3,130-Å radiation is absorbed by CH_2CO?

10-5   Sketch the absorption and fluorescence spectra of a molecule with an $S_0{}^0-S_1{}^0$ separation of 30,000 cm^{-1}. Assume that the fluorescence excitation is produced by 35,000-cm^{-1} radiation and that vibrational relaxation is much faster than fluorescence so that whereas all absorption occurs from $S_0{}^0$, all fluorescence occurs from $S_1{}^0$.

10-6   The photosynthesis efficiency of a strain of algae was measured by irradiating for 1,000 sec with an absorbed intensity of 10 W and an average wavelength of 5,500 Å. The yield of O_2 was 5.75×10^{-3} moles. Calculate the quantum yield of O_2 formation.

10-7   At temperatures below the range in which the thermal H_2-Br_2 reaction proceeds at a measurable rate the photochemical reaction $H_2 + Br_2 + h\nu \rightarrow 2$ HBr proceeds with high quantum yield when the system is irradiated with light absorbed by Br_2 (3,500–5,500 Å). The rate law is $R = aI^{\frac{1}{2}}[H_2](1 + b[HBr]/[Br_2])^{-1}$, where I is the intensity absorbed. Reconcile this rate law with the mechanism for the thermal reaction (Exercise 8-1) and show how the rate of recombination of Br atoms can be inferred from the rate constants of the thermal reactions and the quantum yield of HBr.

10-8   A ^{210}Po source with an activity of 10^9 disintegrations sec^{-1} into a 1-cm^2 area provides 5.3 MeV α particles with average LET in water of 136 keV μ^{-1}. (a) What is the average range of the particles? (b) What intensity (watts) source of 200 keV x rays with LET 1.8 keV μ^{-1} is required to give the same dose rate to a cubic 1.0-cm^3 water sample as the ^{210}Po source? (c) How much H_2O_2 ($G = 3.6$) will be formed after 10 min irradiation?

10-9   One type of nuclear reactor uses D_2O both as neutron moderator (thermalizer) and as the working fluid for a heat engine converting nuclear energy into mechanical work. How much D_2 would a reactor of this type generate per hour if the nuclear power absorbed by the D_2O is 100 kW and $G_{D_2} = 1.7$? Copper salts are used as catalysts to suppress G_{D_2} in such reactors.

10-10   The minimum encounter energy for the reaction $D + H_2 \rightarrow HD + H$ to have non-vanishing cross section was found to be 7.7 kcal. What wavelength corresponds to the 7.7 kcal threshold for D atoms produced by photolysis of DI? The bond dissociation energy of DI is $D_0 = 71.4$ kcal.

10-11   Iodine isotope 128 results from thermal neutron irradiation of the stable isotope 127 according to $n + {}^{127}I = {}^{128}I + h\nu$. The photon emitted is a γ ray of energy 6.7 MeV. Show that the recoil energy of ^{128}I from the emission is sufficient to break the C—I bond in $C_2H_5I^{128}$ [$D(C_2H_5I) = 51$ kcal]. The momentum of a photon is mc, where m is the mass energy of the photon in the Einstein formula $\varepsilon = mc^2$, and c is the velocity of light equal to 3×10^{10} cm sec^{-1}.

10-12 The nuclear reaction $^{37}Cl + n \rightarrow ^{38}Cl + \gamma$ gives ^{38}Cl atoms with $\varepsilon_{max} = 527$ eV. The reaction ^{38}Cl + DL 2-3 dichlorobutane yields DL 2-3 dichlorobutane (^{38}Cl) + Cl; that is, no *meso* compound is formed so there must be retention of optical configuration in the substitution. Write a mechanism, including a steric representation of the substitution step, for the reaction.

SUPPLEMENTARY REFERENCES

For an introductory discussion of spectroscopic experiments, transition moments, and electronic states of small and large molecules, see

M. W. Hanna, *Quantum Mechanics in Chemistry* (W. A. Benjamin, Inc., New York, 1969), Chapters 4, 7, and 8.

Photochemistry is the subject of several advanced treatises:

J. G. Calvert and J. N. Pitts, Jr., *Photochemistry* (J. Wiley & Sons, Inc., New York, 1966).

K. J. Laidler, *The Chemical Kinetics of Excited States* (Oxford University Press, Oxford, 1955).

A. C. G. Mitchell and M. W. Zemansky, *Resonance Radiation and Excited Atoms* (Cambridge University Press, Cambridge, 1934), reprinted 1961.

W. A. Noyes, Jr. and P. A. Leighton, *The Photochemistry of Gases* (Dover, New York, 1966).

P. Pringsheim, *Fluorescence and Phosphorescence* (Interscience, New York, 1949).

J. Saltiel, "The Mechanisms of Some Photochemical Reactions of Organic Molecules," in *Survey of Progress in Chemistry* (Academic Press, New York, 1964), Vol. 2, p. 239.

Radiation chemistry is described by

J. W. T. Spinks and R. J. Woods, *An Introduction to Radiation Chemistry* (J. Wiley & Sons, Inc., New York, 1964).

P. Ausloos, Ed., *Fundamental Processes of Radiation Chemistry* (J. Wiley & Sons, Inc., New York, 1968).

A review of hot atom chemistry has been given by

R. Wolfgang, "Hot Atom Chemistry," in *Annual Reviews of Physical Chemistry*, Vol. 16, 1965.

Index

281